Spring and No Flowers

Memories of an Austrian Childhood

Albertine Gaur

Elm Bank

Bristol, UK
Portland, OR, USA

First Published in the UK in 2006 by
Elm Bank , an imprint of Intellect Books, PO Box 862, Bristol BS99 1DE, UK

First Published in the USA in 2006 by
Elm Bank, an imprint of Intellect Books, ISBS, 920 NE 58th Ave. Suite 300, Portland, Oregon 97213-3786, USA

A catalogue record for this book is available from the British Library

ISBN 1–84150–943–4
Consulting Editor: Keith Cameron
Cover Design: Gabriel Solomons
Cover illustration from a woodcut by Raimund Zotl.

Printed and bound in Great Britain by 4edge, UK.

Contents

The Time in the Garden

The first thing I remember clearly is a dream. The dream came at regular intervals when I was about two or three years old. Not later. The dream never varied. I was in Aunt Paula's room, it was late in the afternoon and the sun was shining. There was a big wardrobe full of beautiful clothes. Aunt Paula opened the door and allowed me to look at them. I remember how I reached out and spread the skirts of the dresses, felt the fine texture of the material, admired the beautiful colours. We did not talk. Talk was something I had not yet fully mastered. We communicated our delight entirely without words. These looking-at-her-clothes seemed to be a well-established ritual, known only to Aunt Paula and myself. Eventually Aunt Paula closed the door of the wardrobe. I clearly remember standing there, with my back to the room, Aunt Paula no longer within my vision, the top of my head just touching the door handle. And then everything changed. Froze. The light, so beautifully golden, took on a deadly white, colourless quality. I took a deep breath and at this moment I knew that if I touched the door handle I would breath out. And then I would begin to scream and go mad. I had a full and absolute comprehension of the meaning of madness. The unspeakable, irrevocable isolation it would bring about. Sanity was only a thin wall of glass, which the scream would scatter. I would pass through it into another dimension, another reality. I thought, but I am only a child, I should not know this, it is too early, something has gone wrong. The grown-ups will punish me if they find out I already know about madness. And at this moment the dream would simply fade away and I, presumably, woke up. But this is something I do not remember.

I was not frightened of the dream in the daytime. If I remembered it at all, I remembered it as something I did not fully understand, and that was therefore of little consequence. But every night the dream returned. Then I also remembered that I had been through it before. There was a particular type of recognition, guilty recognition. Only, as far as I could see, there was no reason for guilt. Just disbelief and shock at my powers of recognition. I realized that my understanding in the dream outstripped my waking capabilities, that in my dream I was transcending the prescribed dimension of my existence.

I have sometimes wondered what could have provoked this wholly premature, abstract recognition of madness. On the surface, our household was secure. The place where I was born was a small baroque town, forty miles west of Vienna. We lived at its southern edge in a rambling house with a large garden. On a clear day one could see mountain ranges from the first floor windows at the back of the house. They were just a gentle blue line in the distance. When I was born my parents, my mother's parents and my mother's younger sister, Paula, lived with us. Not an isolated nuclear, but an extended family, one could almost say a joint family. Such families were in the 1930s not uncommon in Austria. Indeed, as I was to find out, a good many of my school friends lived in a similar way. But ours was not supposed to be a permanent arrangement. In a way it was I who had caused it - or so my mother liked to put it.

My mother was the dominant figure of my childhood, not emotionally, but factually. Her dominance was not so much based on love, but on the fact that she was able to claim exclusive ownership of my person. There was an element of revenge in this claim. She had not wanted me to be there in the first place, now that I had arrived, unasked and at the most inconvenient moment, she wanted to make sure I was kept in my place. I was to be restricted, prevented from causing more trouble. There might even have been a secret element of fear in her attitude towards me. A suspicion that, if not properly watched, I might become her executioner. (As subsequent events proved, her suspicions were perfectly justified. The way we often help to create a situation we most fear by our efforts to prevent it).

Mama's (as I called her) family had originally come from the German speaking part of Czechoslovakia. Her grandmother had been the only daughter of an impoverished (very impoverished) minor nobleman. Hapsburg Austria was flooded with noblemen and noblewomen. Unlike in Britain, there was no clear right of progenitor; all sons and daughters inherited the title. In addition, there was the nobility by merit, who were usually more concerned with their dignity than those who had inherited it. When great-grandmother (her name seems to have been Theresa) was eighteen years old, her own mother died and she was sent to stay with relatives in Vienna. From the fragments of conversation I picked up here and there (none of this was even openly discussed in front of me), the relatives seem to have occupied some minor position at the Imperial Court, and the young girl from the provinces was soon enveloped in the social life of 19th century Vienna. Eventually she fell in love with an officer of the Imperial army and it was there that her troubles started. Officers of the Imperial army were not expected to marry young, their wives had to come either from the right background or, at least, provide a substantial dowry that would enable the couple to live according to their station. Even Jewish women were accepted. Vienna had always been thoughtlessly anti-Semitic but, providing the money was right, and the girl converted to Christianity, the marriage could take place. Whatever else, great-grandmother could not produce a dowry. As to her social station I am not quite sure, but the relatives, in whose house she lived, were certainly not of the right social level. Otherwise her (I think it was some kind of) uncle would not have had to work for money. Whatever the job, at the end of the 19th century 'working for money' always implied lower rank. I am not at all sure what happened. It does not seem that her lover actually deserted her, but his family behaved in a fashion, which so wounded her pride that she wrote to her father, informing him, she would return home and consent to any marriage he saw fit to arrange. Was she pregnant? I do not know. Though I had already developed considerable talent for hearing things I was not supposed to hear, this particular fact would certainly have been beyond my comprehension. The husband her father chose, rather quickly, seemed to have been quite rich (there were apparently a good many debts in great-grandmother's family), much older, a widower, an alcoholic, and there were veiled indications that he had once been 'ill'.

I have only ever seen two pictures of my great-grandmother. One was taken during her stay in Vienna. It showed a slim girl in an enormous crinoline. A tiny waist and soft coloured eyes blazing with light. The other was that of women with three small children. A frozen face, blank eyes. After her father had died, she took her three surviving children and left her husband, a move that immediately made her a social outcast. For the rest of her life she lived in her father's house, saw nobody, took no interest in her children or in anything else. My grandmother's older sister became the head of the household - a very poor household. Great-grandmother used to walk through the house at night with a burning candle in her hand, looking for her lost lover. I was impressed by her constancy, her power to shape life into a coherent story (a Schnitzler story?), divide the essential from the peripheral. Her proud disregard for facts, I admired her like a beautiful poem, a perfect sculpture. I never looked upon her as a possible role model. I knew that even if I tried, my life could never be like hers.

Great-grandmother was entirely indifferent to her children. They were part of the factual world that no longer touched her. This indifference towards children was handed down through my Grandmother to Mama. Her three children, each in his or her own way, carried the taint of her obsession. Great-grandmother died when I was about three years of age. Neither Grandmother nor Mama went to her funeral. Nor did Grandmother's brother who lived somewhere in America. I remember that there was talk that he too had been in the army and that he had left because of gambling debts. (Many years later, when I was a student in Vienna, I happened by chance to come across a woman who had known Grandmother and her brother when both of them had stayed in Vienna. She hinted, maliciously, and therefore not totally trustworthily, that the reason for grand-uncle's sudden departure had been a scandalous suggestion of incest.)

Mama mistook manners for emotions. I had to say: 'Good morning!' individually to each member of the family, curtsy whenever I received a gift. If the gift came from Grandmother, Mama or one of their women friends, I had to kiss their hand. Any omission to say 'Please' or 'Thank you' at appropriate, and sometimes quite far-fetched occasions, brought swift and mostly physical punishment. There was also a rigid set of verbal rules. The most heinous offence was to use words like 'I want' or 'I don't want'. A child, I was told, was not entitled to have a will of its own, a child did not want things but asked for them - politely. Before going to bed I had to kiss everybody and say, 'Good night and sleep well.' The amazing thing was, not that Mama insisted on the strict oberservation of these rules, but that she truly thought they were proof of affection from my side. As long as etiquette was observed, all was well. Just as she was convinced that any lack of proper affection could immediately be corrected by a slap across my face.

I knew open rebellion was completely pointless. So was an appeal for understanding and sympathy. The grown-ups would under no circumstances make concessions, or side with me. Their rules were laws against which there was no court of appeal. All I could do was outwit them, modify the legislation and then stick to my version, just as obstinately and rigidly as they insisted on their right to make laws. It was guerrilla warfare, which solved no problems, but allowed at least some measure of self-respect. I would dutifully kiss each family member before going to bed, but insist on doing it in a certain order, an order I had devised myself. The first kiss went to Nick (my Alsatian dog), the second to Grandfather, the third to Mama, the fourth to Papa (if he was there which did not happen very often), and the last to Grandmother. Grandmother would often complain that she came last, Mama would say bitterly, 'You like your dog better than your own mother.' I never replied and I never changed the order. I sometimes even felt a kind of contemptuous pity for Mama. She may have been a grown-up and powerful enough to make me behave in a certain way, but I already suspected that you could not force love, and evening after evening I was giving them a demonstration of this fact.

Until I went to school I had no contact with other children. Mama's best friend did not have (and did not want) children. As a matter of fact, I cannot remember anybody ever bringing children into the house. When grown-up visitors came I had to curtsy and immediately withdraw to my room. Just opposite our house was a large green lawn, surrounded by birch trees. Some children, living in the neighbourhood, would play there on occasions. Most of them were chaperoned, but some of them were allowed to play on their own. Mama and Grandmother referred to them, contemptuously, as 'street children', a term that implied not so much moral but social inferiority. In their joint opinion there was no need for me to play with other children, or to be taken to a park. I had a large garden all to myself, besides, other children would only teach me bad manners. 'This child has everything she needs,' was a phrase often used by them. To show any signs of unhappiness or frustration was considered a gross act of ungratefulness on my part.

So I stayed within the confines of my garden and invented my own games. I did not really resent my confinement. One day, I knew, I would be grown up and free to go. It was only a question of time. And time was on my side.

There are two games I remember clearly. One was that of 'being queen'. Grandmother could, if she chose, be a prolific storyteller. If I felt like it, I would bring a storybook and if she was in the right mood, she would read to me - providing I promised not to interrupt her. She was very shortsighted (and too vain to wear glasses). At such times a certain bond developed between us, she liked to talk and I liked to listen. I soon began to realize that fact and fiction were not basically different. Following the pictures, she would begin to tell the story slowly and at one point, the tale in the book became intermixed with reminiscences of her own, or her mother's life in Vienna. I knew about the queen in Snow White, just as I knew about the beautiful remote Empress who ruled the Court in Vienna when great-grandmother had still been a young girl. I vaguely began to understand the connection between power and isolation, and I began to see my own isolation not as a deprivation, but as a mark of distinction. I accepted that in order to be different, one had to be removed from everyday happiness (the children opposite the house on the green lawn surrounded by birch trees, laughing, playing, running to their mothers, being hugged, lifted up, kissed). I also accepted the importance of ritual to divide the ordinary from the extraordinary. One Christmas I had been given a little red-lacquered desk with a chair to match. It stood in my room by the window and I used to sit there and look at my picture books and the drawings I had made. Whenever, and as far as I remember, for no particular reason, the time had come to be queen again, I would pull the desk into the middle of the room and put the chair on top of it. Then I would lay an old embroidered shawl that belonged to Grandmother over my shoulders, cover my head with a piece of old lace, and climb on top of the chair. A crown cut from gold paper was always in my desk drawer, together with an old wooden spoon I had elaborately decorated with coloured pencil drawings. Then I would sit on the chair, dressed in my royal regalia, holding the sceptre. I was usually not given to sitting still. Mealtimes were frequently interrupted by slaps across my face for fidgeting. But whenever I was queen, I would sit perfectly still, sometimes for hours. If somebody called for me, I would ignore it (something I would normally not have dared to do), and if a grown-up burst into my room to find the reason for my obstinate silence, I would say, perfectly polite: 'I am sorry, I cannot talk to you, I am queen.' I was never punished for this act of defiance. Even Mama would laugh, close the door, and I would sometimes hear her say in exasperation, but also with a tinge of pride: 'I don't know from where that child gets those extraordinary ideas.'

The other favourite game, I remember, was 'marrying my Grandfather'. I would drape myself into the same embroidered shawl, take my favourite teddy bear, and go to Grandfather's room. 'We are getting married,' I would announce gravely, and just as gravely he would rise from his chair (Grandfather never made fun of me) and we would go in search of Grandmother. She would grumble at being interrupted, but eventually she would relent, and with the teddy bear serving as witness, she would perform the appropriate ceremony. I cannot remember what the ceremony was, I only know that it was a fixed set of actions I had myself invented, and in the end she would join our hands together and pronounce us man and wife. This was the end of the game. Grandfather would retire to his study, Grandmother would return to whatever she had been doing before, and I would go my separate way. I knew the ceremony was not for real, but it served as a sort of preparation. It reassured me, that once I was old enough, I would really marry my Grandfather. Nobody ever tried to stop this game, or tried to explain to me the impossibility, or even the implication, of my action. Silently, they condoned the concept of incest.

I have (apart from the dream) no definite recollections before the age of two. Neither of incidents, people, impressions or emotions. Yet those four people, so oddly joined together for such diverse reasons, their characters, their past histories, their interactions, must in some way have influenced me.

I do not remember Aunt Paula except as a smiling figure moving through my dream in a circle of sunlight. She was a year younger than Mama, or, to be precise, exactly thirteen months. (Grandmother often referred to the fact that she had borne two children within thirteen months as one of the greatest hardships of her life. As a matter of fact, she so dramatized it, that for a long time I thought that this was so dreadful a fate that it had only ever befallen my Grandmother). Aunt Paula seemed to have been cheerful, vivacious, affectionate and warm hearted: all qualities Mama lacked. I still have a photograph that shows the two sisters when Mama was just twenty years old. Mama's face is elongated, oval, with a finely chiselled nose, a small delicate mouth and eyes slanting downwards at the outer corners. It is an entirely Gothic face, like those on the figures guarding the great medieval cathedrals. Aunt Paula's face seems to come straight from a Schnitzler novel. A full, generous mouth, round cheeks and eyes dancing with mischievous laughter. It seems that, from childhood on, Mama had been the obedient, well-behaved, reliable 'good girl'. Aunt Paula broke all the rules, got herself into trouble and was loved by everybody - including Grandmother who normally considered children the greatest curse in a woman's life. Mama was industrious, intelligent, with a flair for music and needlework. Aunt Paula did badly at school, came home in torn dresses, broke Grandmother's favourite pieces of Bohemian glass, and then fling her arms around her neck and cry: 'I am sorry, I shall never do it again.' Mama, incapable of demonstrating emotions, would stand by and eventually be scolded for not having taken better care of her sister. 'You are a year older,' Grandmother would say, neatly shifting the blame from Aunt Paula to Mama.

In the same year the photograph was taken, Aunt Paula, suddenly and without asking permission, joined a sports club. Grandmother, hostile to independent actions and probably jealous (in her youth there had been no such opportunities for young unmarried girls) immediately forbade it. But Grandfather used his power of veto and said, yes, it was an excellent idea providing Mama joined too. From this point onwards, Mama seems to have been entrusted with the responsibility of acting as chaperon for her younger sister. I don't think Grandmother approved of this arrangement. At this stage she still looked much better than either of her daughters, and her sharp wit had not yet disintegrated into maliciousness. Nobody, not even Grandfather, thought for a moment, that Mama might need a chaperon too. In later life, Mama used to complain that her duties as chaperon had greatly spoiled her own chances - she never said what those chances might have been. But she applied herself industriously, and I can not help feeling, enthusiastically, to her task. Aunt Paula was very popular and there were a number of 'unsuitable' attachments. Mama considered it her duty to break them up. She later complained, that there had not been a single dance, a single ball, she enjoyed, because she had to spend all her time worrying about her sister. There seem to have been quite a few occasions when Aunt Paula gave her the slip and returned home alone and much later then Mama, and Mama claimed that for this too she had been blamed. However at the age of twenty-three, Mama suddenly got engaged. The ugly sister had finally triumphed over the fair princess.

(I must, of course, make it clear, that all these stories were never told to me directly. I worked them out from snatches of conversation I overheard when I came into a room, from the way Grandmother and Mama talked whenever I was busy with something else. They were convinced I would neither listen nor understand. Open questions on my part were usually answered with silence, an indication that I had strayed across the permitted line. In consequence, I developed an instinctive talent for entering a room in the middle of a confidential discussion, or look totally absorbed in what I did while a conversation took place. I probably did not understand everything straight away. But eventually, as the stories continued to be repeated, I began to work out my own version.)

For a long while, I thought that, by comparison to Mama's family, Papa's was reassuringly ordinary and uncomplicated. But as time passed I began to harbour suspicions. In fact here too, I knew much less than I thought. Papa's parents died during the First World War. In 1917, when he was just old enough, he was drafted (or volunteered, I am not quite sure which) into the army. Shortly before the end of the war, he was taken prisoner on the Italian front and not released until some years later. When he reached home, he found that his parents had died and that, at the age of twenty-two, he had no proper qualifications, no profession and virtually no money.

Papa's sister Maria was (I think) about six years older. She had for many years been desperately in love with a young man whose family lived, since generations, on an estate in southern Lower Austria. They were a respectable, old established family but very short of money. Papa's family could claim no social distinctions whatsoever, but they owned several hotels in the mountains, and they were prosperous. Both families were anxious for the marriage to take place. But Aunt Maria's devotion was not returned. The young man was greatly in love with a village girl who had already borne him two children. This by itself did not cause much concern (Austrian society has never been puritanical), but what infuriated his parents was the fact that he actually wanted to marry her. When, after the end of the war, financial disaster swept through Austria, Aunt Maria, with her parents dead and her only brother declared missing, assumed financial control over the family affairs. She sold the family property, put Papa's share into the bank and used her half as a dowry. This time she won, but it was a bitter victory. Her husband never touched her. He remained entirely loyal to the mother of his children, and a good deal of Aunt Maria's money went towards his illegitimate family, especially after his parents had died. Ten years after the marriage he was killed during a thunderstorm, while walking back from his mistress' home. Aunt Maria, by then in her late thirties and still a virgin, remarried within a year to a man who was politely referred to as a 'farm manager' but who was actually hardly more than a farm hand. They had one son who was blind. All her life Aunt Maria had great feelings of guilt towards her brother. The money she had put into the bank (or wherever - I have no idea where people put their money in those days) was, by the time Papa returned from the prison camp, just enough to buy a new suit. I have never heard Papa say an unkind word about Aunt Maria. He was, in fact, extremely fond of her. Whenever we visited her she would embrace him, burst into tears and say, 'Please forgive me, I am so sorry, God has punished me already.' Papa would start teasing her until she began to laugh and wipe away her tears. I used to be absolutely fascinated by this performance but when I asked Mama what it was all about, she would purse her lips and say I was too young to understand. Finally I picked up enough courage to ask Aunt Maria and she simply told me.

At the time of the engagement Papa was already in his thirties. He was just beginning to establish himself in business. Then (as now) there is no such thing as Free Trade in Austria. To establish oneself in an independent business means years of apprenticeship, working as a partner, and only at the end the possibility, if not sabotaged by those already in place, to become a 'master', and with it (perhaps) the permission to become independent. All this is rather expensive. At each step there are fees to be paid. Though Papa had but reached the partner stage (I found this out only over the years) he was, for one or another reason, rather successful with women. I expect Mama's initial appeal was her coldness, a coldness he probably mistook for virtue. Mama seemed to have remained cold (and virtuous) and for four long years the engagement dragged on. Mama refused, on the basis of various excuses, to set a date for the wedding. Eventually Papa got her into bed and she immediately became pregnant. Mama was absolutely demented with sheer fury. The very last thing she wanted was a child. She railed against the injustice of the situation. She had not wanted sex, she had not enjoyed it, and as a matter of fact she could do nicely without it for the rest of her life. Papa had promised her there would be no child at least for five years after the marriage. He had cheated her, he had assured her he knew what he was doing and what had he done? Made her pregnant like an illiterate peasant girl the very first time. She tried, assisted by Grandmother (who was appalled at the idea of becoming a grandmother, having never even reconciled herself to being a mother), all the known home remedies: scalding hot baths, alcohol, and eventually large doses of quinine. The latter brought about violent bouts of vomiting, a sickness that did not leave her for the rest of her pregnancy. But I was a survivor from the start. I simply refused to be killed. A month after the discovery, Mama and Papa were married in the largest church in town. I have a photograph taken at the wedding breakfast. It shows Mama with a martyred expression in a white suit, a cloche hat with a garland of tiny white felt flowers, and the whiff of a white veil.

Shortly before all this took place, my grandparents had moved from somewhere in town to the house in which I was born. Mama had agreed to the wedding on one condition: for at least a couple of years, she and Papa would live with her parents. In her state of health she could not possibly be expected to run a household. She was constantly sick (I was told this for the rest of her life) and she had no idea what to do with a newly born infant. If Papa did not agree she would go to Vienna, have an illegal abortion and never speak to him again. It is possible that Papa had created the position on purpose to make her marry him. Many things point in this direction. His complete devotion and loyalty to her, the way he indulged all her whims, as if he had done her, on purpose, some irreparable harm. Most telling perhaps was his attitude towards me. In fact, I am not sure he even had an attitude. I was just there, tolerated like Nick my Alsatian, but never indulged. My attitude towards him was much the same. I had no particular feelings for him, or about him. I accepted his presence, or (more often) absence without question.

Some two years after I was born Aunt Paula died. The official reason for her death was blood poisoning. But for as long as I can remember a general air of evasion, unease and guilt surrounded the subject of her death. Once I heard Mama say, in one of those whispered conversations that immediately stopped whenever I entered a room: 'All that blood, I shall never forget all that blood.' When outsiders touched upon the subject, Grandmother's mouth would tighten and she would say: 'It was the doctor's fault, he did not come quickly enough.' It seems strange that personally I remembered absolutely nothing about so dramatic an event.

Aunt Paula seems to have died within hours, in fact before she could be moved to a hospital. There had been no previous illness. But there had been a drama of some kind. Mama had discovered another 'unsuitable' attachment, so unsuitable that Grandfather, not normally given to such intervention, had gone to see the man and placed Aunt Paula under house arrest. The implications I picked up from several overheard conversations, was that the man was married. But I cannot believe that this alone would have provoked so violent a reaction. Grandfather was neither religious nor puritanical. I have other suspicions but I cannot prove them. The name mentioned in connection with this affair was the same as that of a man who later became a leading Nazi in Austria. I have no proof that this was the same man. But it would have gone far to explain Grandfather's reaction. Grandfather had all his live been an active and convinced Socialist, given the choice between Hitler and Stalin, he would, without hesitation, have chosen Stalin.

Violence and secrecy seem to have surrounded the third year of my life. There was the forever-unexplained refusal of Grandmother to attend her own mother's funeral. There was Aunt Paula's mysterious death. Almost immediately afterwards Mama became pregnant for the second time. This time she did not resort to home remedies. A friend of Papa's, who was a doctor, performed the (then still strictly illegal) operation. Shortly afterwards my mind seems to have started registering events. Life was no longer a tapestry where everything existed simultaneously but one long interwoven string of events.

What then were those first things I remember? Into what kind of world did I emerge? What were the first impressions that produced concrete and lasting memories? There was Mama. I soon learned that as far as I was concerned she held absolute power. Only she gave me orders, only she punished me. She never kissed or cuddled me, she never actually played with me. (I once heard her say to Grandmother: 'All this kissing and cuddling! When they are older they will want it from a man.') I got my toys, my instructions - after that, I was expected to get on with life, quietly and without causing unnecessary trouble. I had a teddy bear in a shade of faded lilac. Despite a set of other stuffed animals, more elaborate and certainly more beautiful, he remained my favourite until I went to school. I disliked my dolls. I never played with them, dressed them, talked to them. The toy pram was used to store picture books and toy animals but I never pushed it round the garden. If asked about my latest doll, I would say: 'She has been naughty, she is kneeling in a corner.' This was indeed the first thing I did with any new doll: make her kneel in a corner with her face to the wall, just as I was made to kneel whenever I had committed, what Mama considered, an especially heinous crime. I had no actual ill feelings towards my dolls, they somehow embarrassed me, they were something I preferred not to touch. On one birthday a friend of Mama's brought me a toy monkey. I promptly began to scream and nothing, no persuasion, no threat, had any effect. Eventually the monkey was removed from my sight. I was not actually frightened. I felt the monkey was some sort of travesty, a deformed human being, and the feeling it evoked was more horrible than actual fear. It was a feeling of being suddenly totally vulnerable, exposed to the threat of infection that might transform me too. I have from earliest childhood always loved animals, but whenever we went to a Zoo, I would never go near the monkey cage, but shut my eyes and keep them firmly closed until we had safely passed it.

I still find it difficult to say what my feelings were towards Mama. I can only assemble impressions, incidents, reactions, but they do not add up, they do not really provide any answer. I certainly considered her beautiful. Especially when she was dressed up to go out with Papa. On such occasions she seemed to change into another person. I would sit at the end of their bed and watch her go through the ritual of transformation. The curling tongue, the rouge, the lipstick, the powder, the scent. I even remember some of her dresses. There was an evening dress in jade green that fell straight from the hips before it opened into two large frills that did not quite touch the floor. And a black, flimsy georgette dress, decorated with red poppies. I felt more and more miserable at the thought that just when she was looking her best, she would leave me. I found it extremely sad and unjust, and I always hoped that this time she would take me with her. When eventually Mama had finished and reached for her evening bag (there was one made of a silver net) or her hat, depending on where she was going, I would burst into tears. 'Please Mama, take me with you,' I would wail, 'you are so beautiful.' Mama would laugh, flattered by my admiration, but simply held out her cheek to receive the obligatory good-bye kiss. In those moments I absolutely hated Papa. He played so little part in our lives and just when I found Mama most desirable, he would claim her and take her away. It usually took some stern words to persuade me to give him his good-bye kiss. After they had left, I would curl up in a corner of my room, my arms around Nick, my face buried in his fur, and quietly cry until Grandmother called for supper.

Whenever Mama and Papa went to a cinema they would leave me a programme. Those programmes were amongst my most prized possessions. I kept them carefully, and when I was alone, I would look at them. Sometimes I persuaded Grandmother to give me a version of the story, or read out the text. In this way, I became familiar with most of the great cinema beauties of the thirties, from Jean Harlow to Greta Garbo. I loved to look at the pictures, but there was never any attempt at identification. I did not think of beauty as something achievable. I was completely unaware of my physical appearance. 'Beauty' and 'ugliness' were terms I did not associate with myself. They belonged to the outside world. They were something at best observed or (as in the case of the monkey, given to me by Mama's friend) violently rejected.

Nobody ever talked to me about Aunt Paula. But every Saturday afternoon we went to visit her grave. Grandmother not only insisted on these visits, she also insisted on walking to the graveyard that was right at the other side of the town, a good hours walk from our home. Mama had to accompany her. She resented it bitterly and once I heard her say to Papa: 'She would have preferred it, if it had been I.' And I think she was right. Without ever openly saying it, Grandmother made it quite clear that Mama's death would have been less of a loss to her. Mama never dared to refuse to accompany her; in a way her domination over me was an exact replica of the domination Grandmother extended over her. I was Mama's chance to feel grown up. Just as Grandmother insisted that she should come with her, so Mama insisted that I should come too. As a matter of fact, I did not mind at all. At this stage in my life, death had no special meaning for me. People simply went away, just as Mama went away with Papa in the evening to go to the cinema, the theatre or a party. The meaning of that 'never to come back again' had not yet impressed itself on my mind. Besides, Aunt Paula was just a shadowy figure moving through a dream, the pleasant part of a dream.

The graveyard was on top of a gentle slope. Once we had passed the tunnel under the railway station there were no houses (except for an old inn on top of the hill), only fields. We entered the graveyard through a large, ornate wrought iron gate. Inside there was a chapel to the left. The chapel consisted of one bare room with a cross, very unlike the baroque churches in town. Opposite the entrance was another door that led to a long corridor. In the corridor, behind glass windows, the dead were laid in state for the funeral. I was absolutely fascinated by this corridor. Saturday after Saturday I asked Mama's permission to go through the forbidden door and look at the dead. Permission was always refused. 'This child is absolutely morbid,' Mama would say. 'I just don't want her to get hysterical.' Since I did not know what morbid meant and since I saw no reason why I should become hysterical, I found the refusal totally unreasonable. But I did not argue. It was the privilege of the grown-ups to be unreasonable.

We always bought flowers at the gate. White flowers: roses, tulips, narcissus and, in autumn, huge white chrysanthemums. After this Grandmother would relapse into silence. 'Work herself into a suitable state,' as Mama put it, when relating our outing to Papa. Once we reached Aunt Paula's grave, Grandmother would begin to cry, quietly and softly, murmuring 'My poor, beautiful daughter,' and other endearments, which, I could feel, annoyed Mama immensely.

It was my job to change the flowers. I would take the containers that stood inside two marble vases on both sides of the gravestone and carry them, together with the fresh flowers, to a little clearing nearby. Our family grave was in a part of the graveyard covered by a pine forest. The trees stood tall and slim, filtering the sunlight in golden, diagonal stripes. The light barely touched the ground covered by a thick carpet of needles. It was extremely quiet in this place. Sounds echoed long after they had vanished. A cone falling to the ground, a broken branch, or, very rarely, the sound of a bird, had an almost shattering impact. I would throw the dead flowers into a large open ditch specially designed for this purpose, and refill the two containers with water from a little hand pump. Up to this day the smell of stagnant water and dead flowers instantly invokes memories of those Saturday afternoon visits. Just as I cannot look at white chrysanthemums without seeing Aunt Paula's grave.

But there is nothing unpleasant about those memories. I did not dislike our weekly visits. Kneeling on the ground beside the fading flowers, I would experience a feeling of extreme tranquility. The grown-ups, at this stage represented by Grandmother and Mama, had temporarily withdrawn from my world and left me to myself. There was freedom as well as imprisonment, a short release from the ordinary world. I was the lost princess in the enchanted forest, every sound had a special meaning that only I understood. Sitting among the graves and yesterday's flowers, the pine trees standing like tall guardians, the shafts of sunlight trying to touch the ground, I had my first feeling of complete and exclusive self-awareness.

Eventually Mama would call out in an aggrieved tone: 'What are you dreaming about?' as if dreaming was one of the most dangerous and undesirable pastimes a child of my age could indulge in. So I would get up and carry the fresh flowers to Aunt Paula's grave. Grandmother would dry her eyes, adjust her hat and say, with a sigh: 'Well, it's too late now,' or 'There is nothing we can do anymore,' in a tone of voice, which, I could see, set Mama's nerves even more on edge. Then we would leave, making our way slowly back, past the war memorial where Grandmother would light a candle. 'For whom?' I would ask Saturday after Saturday and she would reply, 'For my dear brother', in a tone of voice grown-ups adopt when they do not want to hear any more questions. Then we would take a taxi home.

Papa had no place in my world. I remember that sometimes he was away for days. (Nobody told me where he was and I would never ask.) Even if he was not away, by the time I got up in the morning he had left, and when he came back in the evening, he would claim Mama. I knew that somehow he had a right to claim her, and I resented it. I was so exclusively her property, I thought it grossly unfair that she allowed somebody else rights over her time and person. He would bring her flowers, little presents, tell her funny stories, try to amuse her, make her laugh. I used to watch them. I tried to catch their attention. But Papa would say: 'Good grief, girl, you have your mother all day, give her a little rest.' I was again and again outraged by the injustice of this remark. I did not have Mama, she had me. She would watch me, interfere with me, scold me, or teach me a poem (which was fun), or a song (which was no fun at all because I was totally tone deaf). But she would never play with me in a happy, light-hearted manner, kiss me, laugh and chatter the way she laughed and chattered with him. On one occasion I got so furious, I threw my beloved scruffy lilac teddy bear at him and cried: 'You are greedy, you want Mama all to yourself, well, you can take the teddy bear too.' To my extreme chagrin and humiliation they both burst out laughing.

Now, of course, I realize, from fragments of conversation I remember but at the time I did not fully understand, that at this stage Papa was still trying to persuade Mama to leave her parents' house. But she always had different excuses. The most favourite being that she could not leave until Grandmother had got over Aunt Paula's death. But the real reason was that she was simply unable to break away from their protective custody, abandon the role of a young girl and become a grown-up woman. It was easier to be a dutiful daughter than a wife. I am not sure she ever looked on Papa as a husband. He was more the fiancé who one day would take her away - when she was ready. In the meantime she relied on the conspiracy of circumstances: first unwanted motherhood, then duty towards her bereaved mother.

The relationship between Grandmother and Mama was close, though not based on love. It rested on tension, a subtle kind of warfare - they simply were, in the most literary sense of the word, each other's favourite enemies. Mama never talked about Grandmother except in a martyred tone full of suppressed anger. Yet she was unable to disobey her except in a childish, ineffective, furtive way. There were never explained shared secrets, shared guilt and a mutual watchfulness. When my impending birth had become a fact that could no longer be avoided, Grandmother insisted that the coming child would under no circumstances call her Grandmother. She had not wanted to be a mother in the first place, being addressed as Grandmother would only press the point home. In consequence, I have all my life addressed Mama's parents as Father and Mother, and my own parents as Mama and Papa. We all played games designed to defuse the fact of reality by avoiding the blunt edges of their consequences.

There was not too much love lost between Grandmother and myself either. But if we were enemies, we were enemies who respected each other. Grandmother's attempts at domination were more straightforward than Mama's. She ruled by insistence, an acid wit, and, if she chose, humour. She could be extremely entertaining, and there was a certain fearless disregard for people, their opinions and, if necessary, circumstances. Mama would rule entirely through weakness, especially over Papa. Daylong excruciating migraines coupled with an obstinate refusal to take even the simplest painkillers. Nervous, stomach complaints that made her incapable of eating and often caused her to lose weight at an alarming rate. Worst of all were fits of depression that suddenly burst into passionate accusations against anybody who happened to be around at the crucial moment. I learned to recognize the symptoms fairly soon. Mama would stand for hours looking out of the windown, tears glittering in her eyes. It was fatal to ask on such an occasion what was wrong, and equally fatal not to ask. Sooner or later, if I happened to be around, she would accusingly turn on me, claiming that I had ruined her life, made her ill from the first moment, that I did not care for her, that sooner or later I would break her heart, desert her, be ungrateful. (And, of course, that was exactly what I eventually did.)

Somewhere in the story there should now come the old nanny, the devoted nurse, the one predictable figure on whom I could rely. And it is true, there was a nurse, a guardian, an elder brother, somebody who loved me unconditionally, who never let me down, always protected me, whom I could trust just as he trusted me. He was a big ferocious Alsatian, and his name was Nick.

Nick belonged officially to Mama. Papa had given her the puppy, when they first got married, to comfort her during the miseries of pregnancy. When I was born, he was nearly a year old, and he immediately assumed full responsibility for my welfare. If I was left in my pram outside a shop, he would sit immovable like a sentinel besides it, baring his magnificent teeth if an unsuspecting child lover just hesitated a moment to look in my direction. From time to time, he would hoist himself up on one foreleg and peer into the pram, to see if all was well. Somewhere in my subconscious there must be forever the imprint of that huge, shaggy, benevolent wolf's face peering down at me. As far as I was concerned the normal laws did not apply. The mythological symbols had re-arranged themselves for my benefit: the big bad wolf was my special protector.

Until I went to school Nick and I were inseparable. I took my first steps clinging to Nick's fur; if I fell down, he nudged me to get up again. If I hurt myself his long tongue across my face was a reassurance that no real harm had, or could ever befall me. Soon we developed a mutual protection system that was nearly foolproof, and which effectively protected us both from a good deal of physical punishment. He not only protected me from strangers but if need arose, from all other members of the family. Though he was normally devoted to Mama, a slap across my face in his presence usually resulted in a torn dress. He knew, if there was real trouble in store for me, almost before anything happened. If Mama wanted to slap me, she had to be careful not to raise her voice, and go first of all in search of Nick. If she succeeded outwitting him (and she seldom did), she had to entice him into the cellar and lock the door. Only after this had been achieved (at considerable exertion of time and energy) could she tackle the task of rounding me up. I hardly ever cried because of fear or disappointment but to be beaten (usually with the bare hand or if she was especially furious with a carpet beater) I would scream and struggle and scratch. It was the barbaric injustice, which infuriated me - after all she could only beat me because she was bigger! At the first scream Nick would scratch the cellar door and set up a howling that would have done justice to a Siberian wolf. Exhausted by the infernal noise, Mama usually had to give up after the first few slaps. On the other hand, whenever Nick had committed a crime (the most serious ones were to dig out Papa's roses, bite the postman or any other unsuspecting person who had ventured into our garden), he would make straight for me. I immediately recognized the signs and, throwing myself over him, I would start to scream until, unable to draw a breath, I was gasping for air. The grown-ups had to lift me up, shake me, make me drink some water and, on a few occasions, even send for the doctor. Nick and I were absolutely in earnest; we did not put it on. If they had touched Nick in my presence, I was determined to scream until I suffocated, just as he would have killed Mama had she slapped me in front of him.

In sheer self-defense the grown-ups devised other forms of punishment. Instead of inflicting violence, Nick was put on a chain in the garden (which he hated) and I was made to kneel in a corner of the living room with my face against the wall whenever one of us had offended against their arbitrary laws and restrictions. If we were not both punished at the same time, we did our best to reach each other. Once established firmly side by side, the grown-ups might just as well have given in straight away, because neither of us was in the least upset anymore. While Nick's punishment had at least a time limit, I was supposed to apologize and say I would never do it again (whatever it was I had done). This I simply refused to do. I felt I had offended, I was punished, and, as far as I could see, that was enough. I was not prepared to add the humiliation of an apology to the punishment. I had paid the penalty, so why should I apologize? It was a point on which I simply refused to compromise. Sometimes I was still kneeling in my corner, when Papa came home in the evening. He usually tried to persuade me to say I was sorry, but I never relented. Everybody got exasperated with me. Mama would work herself into one of her migraines, but I would not budge. It was eventually Grandfather who put an end to the whole farce by simply coming into the room, and quietly telling me to go to bed. Clutching Nick's fur, we would both depart upstairs.

Grandfather was the remote but ever present centre of gravity in our house. Though he seldom spoke and never issued orders, I knew that nothing could or would happen against his will. In his presence Mama and Grandmother stopped bickering, gossip died down, nobody ever carried tales to him. At mealtimes he sat at the head of the table, which was enough to ensure peace and order.

I did not particularly like food at this stage. Some dishes I found quite tolerable but there were others I found unbearably revolting. I could not bring myself to touch them. Refusal to eat was in Mama's eyes rebellion, and rebellion had to be repressed at all costs. Children did not like or dislike things, they were grateful for what was given to them. I would sometimes try to offer acceptable alternatives, like 'I am not hungry today', or I would promise 'to eat twice as much tomorrow' if I was left off the hook. But that 'you eat what's on your plate' syndrome was one of Mama's favourite power games and nobody was allowed to interfere. It was not just a question to sit in front of my plate indefinitely; all refusals to eat were followed by sharp slaps across my face. At one point she tried threats like 'If you do not eat up you will not be allowed to go into the garden in the afternoon' or 'You will not get any cake today.' I soon learned that one could easily deflect such threats by simply saying, 'I don't want to go into the garden anyway' or 'I don't want cake.' This as a matter of fact was quite true. The revulsion increased with the pressure put on me, I would willingly have promised anything, if I could only avoid the horrible moment when I had to put that thing on my plate into my mouth. I never cried, I just sat there, politely arguing my case and suffering the slaps without any visible change of expression. Grandmother would sometimes say in exasperation: 'This child is like wood, she has no feelings whatsoever.'

Eventually the matter came to a head over a plate of spinach. I hated spinach, even the colour made me feel quite sick. On this particular day Mama got more and more angry. Papa (there by chance), seeing she was working herself into one of her weeklong migraines, for once intervened: 'Can't you see you are making your mother sick? A spoon of spinach won't kill you.' Surrounded by unrelenting grown-ups, I ate the spinach and immediately vomited it back on the plate. Mama rose, her face white with anger, Nick began to bark in the garden - this had the makings of a first rate scene. But Grandfather raised his head and said quietly to Mama: 'This child is not going to be forced to eat food she does not like ever again.' Mama sank back into her chair; I had miraculously been given full absolution. Grandfather would stand by his word. If he said I was safe from this particular peril, I knew I would be safe for all eternity.

After this incident Grandfather began to emerge from his benevolent remoteness and entered my world. He began to teach me things. Mama's teaching was entirely negative. It was always 'Why have you done this when we are trying so hard to bring you up properly?', never 'It would be nice if you did so and so.' Grandfather introduced me to observation, working things out for myself; the first fundamental principles of thought which held the seeds of logic.

He began to take me for walks. At first into the immediate surroundings, later into the mountains. Since our house stood at the southern edge of the town, one had to walk only a short distance before reaching open country. One place became my favourite spot and remained so until I left Austria. It was an old neglected mill, long disused, standing in the middle of a large ground beside a millstream. There is something about neglected houses that exercised (and still does) a special fascination for me. They exist on the border of everyday reality, they could be inhabited by figures from the past, aspirations of the future, the held promises, secrets - everything might happen behind the high walls under the thick overhanging trees. The mill house had a tower on one side, which could clearly be seen from the road. The tower seemed specially designed for me - waiting for me. Later, once I had gone out into the world and conquered it, I would retire into the tower and allow only my most trusted friends to come and stay with me from time to time. Grandfather would listen silently to such announcements. But it was a silence that did not belittle my aspirations. It made me feel it only depended on me to achieve all this, and that Grandfather had confidence in me. While the mill house exercised its fascination throughout the year, there were other spots I remember: one was a spring the other a winter place.

Sometimes we would take a train into the mountains and go for walks that could last a very long time. Those walks taught me invaluable lessons, they taught me one did not give up just because one felt one could not go on any longer. Grandfather would tell me about flowers, birds, trees, how to find directions, tell the time of the day, other places in the woods. He never talked directly about personal things, but by telling me about the world in general he made me feel part of it.

Grandfather never complained, and in this way, he taught me that complaints were ineffective ways of trying to solve problems. Those walks were beautiful. I can still see glimpses of certain scenes, a bend in the road, blue mountain flowers, a little waterfall. But they were long walks and I now realize that Grandfather specially designed them to stretch me to the limit. I would say 'I am thirsty, Grandfather,' but he would either remain silent or continue with the story he was just telling me. I would say 'I am hungry, Grandfather,' and he would reply 'We shall find some food once we have reached our destination.' I would say, 'I am tired', which was usually simply ignored. A few times I would say, 'I am so tired, I can't go any further.' To this he would reply, 'All right, if you really can't go any further, you'd better sit down here. I shall go on and when I come back you will be rested enough to come to the station.' This was an idea that did not appeal to me at all. So I went on, realizing that the fact I felt I could not go any further was no reason whatsoever for giving up. Tentatively I learned that it was possible to stretch oneself. To reach beyond what at first seemed the very limit.

Grandfather also taught me the first elementary sense of history. I learned that each object, each tree, church, farmhouse, each street in our town had existed long before I had set eyes on it. There were stories attached to each of them. People too had stories. Some obvious, some secret. During the war Grandfather had been taken prisoner on the Russian front and sent to Siberia. There he had first of all learned Russian, and then organized an escape. Together with a small party he had started the journey back to the Austro-Hungarian border. This story never failed to fascinate me. It was my first story of proper adventure, an introduction to epic poetry, a ballad told by an eyewitness. It gave me the first inclination of danger, of the fascination of danger. I heard the story so often, I asked for so many details that, even today, I am not quite sure what Grandfather told me and what I imagined or added to it. I never talked about Grandfather's stories to anyone. It was a secret only shared by the two of us.

Our family was not religious. Though we were, like nearly all Austrians, nominally Catholic, neither Grandfather nor Papa ever went to church. Papa did not believe in God; he felt religion was something best left to women. Grandfather had his own arrangements, which I came to understand only much later. Grandmother held a personal grudge against God, making him responsible for Aunt Paula's death, thus sharing her burden of guilt. If the talk turned to Sunday mass, she would say, 'I visit my daughter's grave every Saturday and this is enough church going for me.' Mama would sporadically take me to church on Sundays, but her visits too were not based on any deep religious feelings. She simply liked to get away from the house (and the responsibility of arranging Sunday lunch, which was conveniently left to Grandmother) and meet friends from the days of the Sports Club. Together they would afterwards go to a coffee house and gossip. They would usually stay late and lunch had to wait for us. This made for a chilly atmosphere for the rest of the day. My own feelings about religion were, at this stage, rather ambivalent. Grandfather's influence had taught me to rely on my own resources and not expect divine intervention to see me through. Papa was too remote to influence me with his atheism. I was told the type of Bible stories children are usually told, but they did not move me. I was more impressed by Pontius Pilate, who had power over life and death, and who could pronounce a sentence, which Christ meekly accepted. The truth is I never thought much about it. But I liked the drama of the Catholic mass, the rich costumes, the triumphant sounds of the organ sweeping though the gold glittering nave of the large Baroque church. (I have always like the colour of gold.) The chanting of the priests at the altar in a language I did not understand, and the answering chorus, which gave me a feeling of being, however fleeting, part of a group, all moved by the same exaltation. It was not necessary to understand the exaltation; the excitement lay in being able to feel it. I most definitely did not like the time in the coffee house with Mama's friends, where I had to sit absolutely still, keep quiet and only reply to questions, usually of an infuriating manner. At home nobody talked to me as a child, and the cooing questions about my dolls (which I hated), and about the length and colour of my hair, seemed false and stupid.

In those coffee house visits there arose the first seeds of Mama using me to show off. I could quite easily remember and recite poetry. Somehow she would bring the conversation to a point where one of the women simply had to ask me to recite a poem. I hated doing it. I felt they did not really want to hear me recite poetry at all. Mama had just conned them into listening, showing an interest they did not feel. They would probably make fun of me afterwards. I felt I was being used to make an exhibition of myself for Mama's benefit.

By making the world outside the garden taboo, the grown-ups enclosed me firmly in their own world. There were certain topics, I felt, that were not freely discussed in front of me. I had vague ideas that there was trouble outside our house. One night there was shooting in our garden and Papa told Mama and me to lie on the floor. I knew about the murder of Dollfuss but only because he had been a personal friend of Papa's. I heard the name of Hitler, that Germany was holding some kind of threat over Austria, that some people in Austria secretly belonged to Hitler's party, and that their signs were white knee-high socks and blue cornflowers. Grandfather's godson (or was it a nephew several times removed?) of whom he was very fond, had once been beaten so badly that he had lost an eye, only because he had worn white knee-high socks and a corn flower in his lapel. Yet he and Grandfather remained firm friends. I knew that many people were poor and out of work. I had no idea what our position was, either socially or financially, simply because there was so little outside contact. But I did not think of us as poor. Nobody ever talked about money. Mama seemed to buy whatever she liked. She had drawers full of beautiful materials, which she was going to make into dresses for both of us - but never did.

There was safety and distinction in our seclusion. I felt that whatever happened outside the walls of our garden, the magic circle I was not allowed to cross, had only the importance of another story. It could not really affect us personally. The beggars who sometimes rang the bell (passionately hated by Nick) were only reminders of an outside reality, it could happen to the people out there, but not to us. I was suspended in my childhood, which in those days I could not imagine to end. There was no feeling of the existence of time as continuity, incidents happened side by side, on one level.

The idea of sex did not enter my world in any recognizable form. Since I did not play with boys or girls, I was not perturbed by any suspicions of a possible difference between the sexes. From personal experience men certainly did not seem superior, our household was entirely dominated by women. Papa was a mere appendix. Grandfather's position seemed based on personality and strength, certainly not on gender. I never asked the question after the origin of babies, but I was exceedingly curious where puppies came from. Until one day, whilst walking with Grandfather, we came across a bitch with heavy tits, surrounded by puppies. Suddenly it was obvious to me. 'Did she carry the puppies in her stomach before they were born?' I asked. And Grandfather simply replied: 'Yes.' With this the question was solved and I lost interest.

Since nobody in our house liked children the grown-ups converted me into a powerless little adult. My special retreat, the place I remember most clearly, was the garden. It was the place where for most of the times I was left alone with only Nick for company. The garden was entirely Grandfather's creation. From the terrace at the back of the house one stepped down into a part covered with pebbles which we called the 'court'. At one end of it there was a swing where I used to spend hours, trying to discover new ways of using it. There was an iron chain on one of the supporting poles; this was where Nick was chained if he had committed an especially objectionable crime. I would then sit beside him on the rough ground until the pebbles embedded themselves so deeply into my legs - I thought I would never be able to remove them. From the court, partitioned from the rest of the garden by a low fence, two small gates, each crowned by an arch of pink rambling roses, led to the formal part. The garden was carefully laid out. There were pebble covered pathways, ornamental flowerbeds, and a goldfish pond in the centre. Behind the formal garden was a little orchard with fruit trees: apples, peaches and pears. Behind it lay the strawberry and gooseberry beds I was not allowed to touch before the proper season. Yet they always tasted much better earlier in the year when the fruits were still sour and unripe. I have no memories of the garden ever having looked differently, which meant that Grandfather must have laid it out soon after the family moved into the new house. Papa had no special interest in the garden apart from two rose beds, where he created proper roses out of wild bushes whenever he found time. In winter they were covered by a mass of yellow straw to protect them from the frost. There were two small lawns in the garden surrounded by ornamental borders filled with low growing flowers. I liked to lie in the grass in summer and watch the big white summer clouds changing their shape as they sailed by. There were dragons and fishes and rams and at times a beautiful face looking down at me over the border of eternity. Lying on my back, with my head on Nick's body, I had glimpses of perfect harmony, of being part of, and belonging to, something more permanent than mere people. In winter one could built castles in the garden out of snow, form little mountains from which to slide down, or tunnels through which one could crawl, imagining that at the end of a tunnel one might suddenly emerge into a world unlike anything one had ever seen before. It was in the garden where I lived between reality and imagination.

For some reason, a minor incident, that happened when I was around five years old, left a deep impression. It happened during a walk on Sunday afternoon. Sunday was (is?) the day for the obligatory *Ausflug* in Austria. The *Ausflug* could either be a day trip into the mountains, or an afternoon's walk through the surrounding countryside. Both ended with some refreshments being taken in some inn - an occasion for Nick to display his fighting abilities. On this particular afternoon the whole family was going for a walk in what was still called the *Kaiserwald*. We were walking up a little rise and I was talking to Grandfather and Papa. I cannot remember what I told them, but I was deeply engrossed and totally unaware of my surroundings. Halfway up the little hill, I looked up and, to my horror, realized that I was talking to a complete stranger, who looked down at me with an amused smile. Confused I looked around and there, on the other side of the gently sloping meadow, stood the whole family, laughing, watching me. The path had forked some way back and, for the last five minutes, I had been walking next to a complete stranger, telling him things that did not concern him. I felt humiliated and betrayed. How could they have possibly allowed me to make such a fool of myself? Even worse, how could they just stand there and laugh at my mistake? The stranger said something, kindly I remember, but I tucked my head between my shoulders and slowly walked across the meadow to join the others. I hoped, they would at least have the decency to ignore the incident, but no, they teased me, joked, and I felt more and more furious. I did not say a single word for the rest of the afternoon and eventually I was accused of sulking and rudeness.

I had none of the usual children's illnesses, perhaps because I simply had no contact with other children. But I clearly remember that from time to time I spent a day in bed, vomiting and with a terrible headache. Mama put this down to having eaten something bad. She had in fact a whole set of anxieties about certain foods: too cold, too hot, a combination of both, not quite fresh enough, too much butter - all of them produced similar symptoms in her, with only one difference: mine were less frequent and lasted for just one day, hers could sometimes last for more than a week.

Papa was perhaps not so much a negative entity but a curious kind of vacuum. He never scolded or punished me. There was no ground for resentment and fear, he just played no part in my life at all. Grandmother talked about him with thinly concealed contempt, as she well might, she was after all the winner and he was the looser. Mama was staying with her; she was not leaving her for the sake of Papa. To Mama she talked about him full of reproach, making it by implication perfectly clear that he did not count, that he was a superfluous addition to the family. Mama never took his side, she never defended him, if at all, and she was rather apt to complain. About what I had no idea, but in some ways he seemed to make demands on her which she considered wholly unjustified. Since I was often just around during those discussions, I heard what the grown-ups said, but I hardly ever understood it. Nevertheless at the end I was always sworn to secrecy. 'You are not to tell Papa,' was the refrain. At this point in my life I found it rather difficult to tell lies. It was quite easy to invent a story and then tell it as a fact. Because of my isolation the borderline between reality and fiction seemed often extraordinary precarious. But if asked a straight question, I usually gave a straight answer. I could not understand why I should not do so. The person in question had after all asked, surely you did not take the trouble to ask, if you did not want to know the facts. Since Papa had seldom anything approaching a personal conversation with me, the order of not telling caused little difficulties. But there were a few occasions when I got into trouble.

There is one memory, which is still not quite clear in my mind. I was lying in Mama's bed next to Papa. Mama must have been somewhere else. I was happily talking away. Telling him something that must have happened the previous day. Until I suddenly realized that right in the middle of my story, there was a large area of taboo, something Grandmother and Mama had said to each other, something I was not supposed to tell. Though usually good at inventions my mind became more and more blank. I knew it should be easy to manoeuvre around the dangerous subject but the very thought of betraying a secret (I have no idea what the secret was), made my speech go slower and slower until, eventually, I began to stammer. This seemed to arouse Papa's suspicion and he began to cross-examine me. I became more and more confused. I had been told not to tell (whatever it was), and to act as an informer, was, as Grandfather had taught me, the dastardliest thing to do. But much as I tried, I simply could not think up a lie quickly enough. Papa kept on and on and eventually I blurted out 'I am not going to tell!' He grew angry and shouted: 'Then get out of the room immediately.' I got out of bed and went to the door. Papa tried to call me back, his voice changed, but it was too late. My indignation at the humiliation, and the disgust at the deceit practised by the grown-ups was too great. On the one hand you were told not to lie, on the other hand you were told not to betray secrets and break promises. And if you followed all their instructions, they were far from pleased, they were in fact furious.

I had no idea what Mama's and Papa's relationship was in those days. How happy or unhappy they were. Papa, I think, still hoped to persuade her to leave her parents' house and set up a home with him. How far he blamed my existence for the unnatural state of affairs, how far he blamed himself, Mama, Grandmother, or even Grandfather who seemed like a remote sage, refusing to give instructions or advice - I have no idea. Even how Grandfather judged the situation. How far he was indifferent or whether he thought Mama and Papa had to decide their own life, that Papa had to have enough strength to take Mama away, and Mama have enough courage to go. One thing I remember is Mama sometimes complaining that, when she was a young girl, she had desperately wanted to go to the music academy in Vienna, but that Grandfather had dashed all her hopes with the simple statement: 'You are not capable of leading an independent life.' She often repeated this to Papa, adding bitterly: 'It was just as if I was not normal. As if I could not look after myself.'

I do not remember any actual quarrels. But I remember one incident that I have never been able to explain to myself. There had been some talk, raised voices between Mama and Papa, perhaps even Grandmother, I am not sure. Suddenly Papa began to shout, but not like somebody in anger, more like somebody in pain. Then Grandfather came out of his study and he and Mama helped Papa upstairs and into bed. Papa was talking excitedly, incoherently; it conveyed something like mental not physical despair. I stood in the bedroom door, clutching my lilac teddy bear. When Grandfather saw me he said: 'Just go!' and I stole down the stairs. Nick was lying in the hall under the staircase and I curled up beside him, still clutching my teddy bear. There was much coming and going, the doctor came with a big bag, Mama's hysterical voice, Grandfather's quiet and with authority. I stayed in my place, keeping well out of sight. I felt everything was out of order, any mistake on my part could tip the scales; hurl us all into an abyss. Papa stayed in bed for some time, there was some talk about his heart not being too good, but I knew it was something more, something different. The incident never repeated itself and eventually everything faded back into its usual shades. Life went on as before. But for a moment I had perceived that Papa was not the indifferent nonentity he had been made out to be, that there was something in him that could break, something beyond the manipulations of Mama and Grandmother. But I did not know what it was.

There were certain places in the house, certain pieces of furniture, which were invested with special meaning. One, I remember clearly, that fascinated me for many years, was a big gold-framed mirror that hung in the hall. It had an elaborate baroque frame, and the glass was of a particular matt veneer, which made everything it reflected look more beautiful. To begin with, I was still too small to see myself in the mirror, but if I jumped up and down, I could, for a moment, see my face. I was entirely without vanity or concern about my physical appearance, but the momentary glimpses of reality outside myself that flickered across the silver screen, fascinated me. There were other mirrors in the house but none held the same fascination.

On another wall hung an old clock. It had heavy weights and a continuous, gently swinging pendulum. Once a week, Grandfather would wind it up, simply by re-adjusting the position of the weights. It was something I rarely failed to watch. It seemed fascinating that one could tell the end of a week simply by the position of those heavy metal objects, and that by adjusting them Grandfather transformed the end of one week into the beginning of the next. I never inquired what would happen if the clock stopped. I had implicit faith in Grandfather. As long as he was there time could not come to a standstill.

There was a particular part of the staircase where, I knew, lived a witch. I never told anybody about it, because I knew nobody would believe me. It was a peril I had to face alone. I was not particularly frightened about it, because I knew, if I went past this place in a certain way, completely noiseless and on tiptoe, the witch was powerless. She was also powerless if a grown-up accompanied me. Trouble only arose, if for some reason, I was sent upstairs to fetch something, or if I was unexpectedly called down. I knew I could not run past this particular place in the staircase, I had to walk very slowly, use the exact amount of time required for the ritual. The grown-up would become more and more annoyed and call out, asking what on earth I was doing and why I could not come down promptly when called. Part of me knew that if they caught me in my deadly serious game, it would be even worse, because they would want to know what I was doing, and I could not tell them. They would either laugh at me or become angry. The only time I could walk past in a normal way was when Nick was with me, then I would firmly close my eyes and walk past holding my breath.

There was another peril. As soon as I lay in bed, I knew that a giant was walking around the house, looking into every window to see if somebody was there. If he found any signs of life, any movement (even breathing could make you visible), he would simply reach into the room with his hand, pull you out and eat you. The giant was exactly the height of the house and his eyes were just above the first floor windows. He went around the house in a definite order that never changed. My room was the last he visited. I imagined him looking into my parents' bedroom (incidentally my grandparents were perfectly safe because their bedroom was on the ground floor), and since there were no shrieks of horror, he had obviously not seen them. And then he came to look into my room. I lay perfectly still, not too frightened because I knew I was capable of outwitting him, it was just important not to move, not to breathe, not to open one's eyes. After the proper time, the giant simply went away and I knew he would not bother me until the next night.

There were other places in the house I still remember. In our kitchen stood a large old-fashioned dresser. If I climbed from a chair on to it, I could lie underneath the overhanging top part that held our special, precious china collection. It was a secret place I loved and where I was perfectly happy to stay for long periods. Secretly, because I would, of course, not have been allowed to climb with my shoes on the dresser and lie there. Grandmother would angrily have reminded me of the basic laws of hygiene, and Mama would have put the whole adventure into the forbidden category of 'dreaming'. Our kitchen was large and on the ground floor. It occupied a corner of the house and there were windows on both sides. Underneath the two corner windows ran a built-in cupboard where (among other things) a large china bowl housed all our sugar cubes. One day, for no reason at all, I got up from my comfortable hiding place, took a little cup of water kept for Nick's bowl and emptied it into the sugar container. It ran through the mound of sugar, softening the harsh edges of the cubes here and there. I became absolutely fascinated. I brought another cup of water, and then another, until at last the whole mound of sugar sank to one even level. I replaced the cup after having careful drained and dried it, and left the kitchen. In the evening there was great consternation and much talk about the strange phenomenon of the sugar cubes having turned into a soggy mass of white sugar. I was asked, just in passing, whether I had put water on it. By the sound of the question I realized that it was just a token query, and that neither Mama nor Grandmother really thought I could have done anything quite so pointless. So I just shook my head and they immediately turned their attention away from me. Dimly I realized, that as long as a crime was absolutely without motive, one stood a good chance of getting away with it.

Another favourite place was the small basement underneath the outside terrace one entered from the garden down a steep narrow staircase. It was kept mainly for storing gardening tools. It was a dark, cool place; the only light came through a tiny window near the ceiling, mostly overgrown by the wild wine that covered the verandah. One could not only close the door, but there was an additional trap door that fitted over the stairs to keep them dry during the rain or heavy winter snowfall. Normally the trap door had to be closed from the outside but it was just possible to close it from underneath. In any case, the closed trap door indicated that the basement was empty. In those days I still had to go to bed between two and four in the afternoon. I found this intolerably irksome and boring. But those two hours were religiously observed. All protests that I was neither tired nor sleepy were ignored by Mama, or countered with a 'That may be as it is but I want to have at least two hours of peace.' Once in my room I was strictly forbidden to do anything but lie in bed. However noiselessly I tried to climb out, Mama always heard me and immediately came up to establish order. Eventually (I think it was when I was about four) I grew more and more frantic to avoid those two hours of unbearable boredom and tried everything possible not to be available at 2 o'clock in the afternoon. At first I tried the trees in the orchard, or hiding behind the gooseberry bushes at the very end of the garden - but to no avail. Eventually I hit on the outdoor basement. Mama did of course eventually discover me there, but it took her nearly an hour and Nick as usual foiled her attempts to punish me.

I was not a pretty little girl. I had exceptionally thick and heavy blond hair that was pulled back, from my much too wide and too high forehead, into two enormous plaits. From one of the rare surviving photographs I can see that it made me look like a head walking on two legs. After my birth Mama had been horrified to find that I had no hair at all and, stunned by the prospect of having produced a hairless daughter, began to rub birch essence into my head. The result was startling. She would later say, not without a certain satisfaction, 'That was typical of you, always extremes, first no hair and then too much.' Washing my hair always ended in tears. I tried to pull away, got shampoo into my eyes and eventually some slaps from Mama's shampoo-covered hand, which produced more screaming. Trying on a dress was another fraught incident. After her marriage, Mama never touched her violin again but she began to make all our dresses. She actually loved to collect materials, but somehow they were hardly ever used. Having a dress made meant endless fittings, standing still for what seemed hours. I had no interest in pretty dresses and usually tore them or made them dirty as soon as I put them on. The fact that I was told by Mama, Papa and even Grandmother (jealous, because she loathed seeing anybody receive special attention apart from her) that there were many little girls who would be happy to have such beautiful clothes, made no impression. I simply was not interested in how I looked. With one exception.

Brassieres. I developed an absolute fascination for them. I just could not understand why I could not have one. Mama had one, Grandmother had one, so why not I? I began to torment Mama to buy me a brassiere but it was always refused with the explanation 'You are too small.' This I had to accept. I was smaller than either she or Grandmother. So there was obviously no point in fighting against it. Until one day I went with Papa into town and saw, in a shop window, a small dummy wearing a miniature brassiere. 'Papa,' I cried, 'look at that brassiere, that one will fit me. Please, Papa, buy me that brassiere.' My outburst caused hilarity among the people around us. Papa and I became involved in a long argument why I could not have the brassiere when the dummy in the window was just as small as I was and could wear one - until I was eventually dragged home.

In Aunt Maria's place none of the normal restrictions applied. Mama left me more or less to myself. She just was not as evident in my life as usual. Now, I wonder, whether she was simply happier there, away from home, from Grandmother, alone with Papa. Most of the time I was left to myself, to Aunt Maria (who I felt truly loved me - though I now cannot for the life of me remember how she looked), the animals I absolutely adored. Mama complained that I constantly stank, that my hair and clothes would reek of cow dung, horses and pigs. At the end of the day I was usually dirty and, since I normally ran around barefoot, my feet had to be scrubbed with a nailbrush. I was never slapped in Aunt Maria's house. Perhaps Mama simply could not be bothered or perhaps she did not want to give the impression of being tough, when Aunt Maria was so obviously emotional and soft. Most of my time I spent in the fields or in the cowsheds. I liked to watch the cows being milked, lean my head against their warm bodies and watch the stream of white milk foam into the wooden pitcher. A few times I tried it myself but without success. The new calves, kept in a special enclosure, were another object of wonder. I used to climb over the railing and sit with them on the soft straw. I was thrilled when I was shown how to wean a calf: put one's arm into a pitcher of milk, the hand into the calf's mouth and make it drink. There was a big black bull kept in a special enclosure. He was supposed to be dangerous and I was strictly forbidden to go near him. Sometimes I would creep into that part of the stable when nobody was around. Eventually I managed to climb into his food crib and stroke him between the horns. The big black beast turned his head in pleasure and stood there absolutely transfixed. A stable boy, who one day discovered me, related the story back to the other farm workers with a kind of possessive pride and they never told anyone. They showed me how to ride bareback whenever the horses were taken to a nearby little pond. I would sit on the big farm horses when they went into the water, just about managing to hang on, feeling perfectly safe. I knew one could trust an animal the way one could never trust a human being.

There was also, I remember, a special pigs' house. The pigs always evoked ambivalent feelings in me. They seemed perfectly ordinary and dull, but then at feeding time, as soon as people came with wooden bowls, they simply transformed themselves. Instead of the occasional sleepy grunt, there was suddenly an infernal noise. All the pigs standing on their hind legs, trying to look over the top of their enclosure, screaming in the most blood curling fashion. There was a big old boar. I had been told to keep well away from him. He was vicious and dangerous and once he had nearly killed one of the farm workers. I was repulsed and fascinated by him. I would climb on the wall of his enclosure and look down at him, wondering what exactly it was that made him so different. Once, balancing around the rim, I lost my footing and fell right on top of him. The boar, which all his life had been treated with the greatest respect, was absolutely stunned. He retreated into the opposite corner, eyeing me doubtfully, waiting for my next move. It was one of those moments when I realized I had gone too far. I was too small to climb out, and so we stayed (what seemed an eternity) just looking at each other. Fortunately somebody happened to come by and quickly pulled me out. As soon as the boar realized I was in retreat, he charged with a terrific scream and hit the corner just below my feet. My adventure was discussed among the farm workers and they made me promise not to behave in such a stupid fashion again. In a way, they seemed actually quite proud of me. I heard one of them say 'That girl is going to do something special, she isn't frightened of anything.' I was proud of being considered special, and frightened because I dimly felt that this put a strange obligation on me to do something - I just had no idea what it could be.

All memories I have of Aunt Maria's house are warm, full of gentle colours, the smell of hay, the soft noses of horses and calves. There were a group of wild ginger cats, which were fed with milk in the cowshed; they did not allow anybody to touch them. Even kittens shot out from underneath your hand, fiercely independent and arrogant. Sometimes Aunt Maria would take me to the hayloft to look for eggs, and we would return to the house with a huge wicker basket full of white eggs. Sometimes she would take me for a walk to visit the young heifers, which during the summer were kept free on a piece of pasture up the hillside. For those walks, we would take red salt with us and as soon we entered the enclosure they would all come running and take the salt from our hands. Sometimes they would lick my face with their rough tongues or try to swallow part of my dress, which always greatly amused me.

Our departure used to fill me with deep misery. Aunt Maria would give me a little present and whisper how I would be back again soon. By the time we reached home and the taxi turned into our road, I would invariably burst into tears and, to Grandmother's disgust, arrive howling at our doorstep. 'I can see you did not miss me,' and 'Aren't you glad to be home again?', she would ask and I would howl even louder and be sent to my room to recover a more suitable demeanour.

I cannot say when exactly the dream began. But I know it was in the year when I had to go to school in the autumn. Sometimes during that year, and it was probably in spring, the dream started. A recurring dream. It came night after night, immediately after I went to sleep. I knew every evening that it would come again. There was no escape, no possible retreat. Yet I could not tell anybody, because I knew they would only laugh and tell me not to be silly. There was no way of avoiding the dream. Only the possibility of delaying going to bed, perhaps for half an hour. But that was no solution.

The dream never varied. It was exactly the same down to the last detail. I was in the living room, Mama was there with some friends, and they were all looking at fashion magazines. I was sitting under the table. The table was round and it was covered by a crocheted white tablecloth. Nick was lying beside me.

And then I knew that the lion was already in the garden. I knew exactly where he was; he was lying under the peach tree. He was a large animal with an enormous head and a huge mane. I knew the lion had come for the sole purpose of killing me. All the people in the room knew about it, but nobody seemed prepared to do anything.

Then I realized that the lion had got up and was slowly moving through the garden, up the steps of the verandah. At this point Nick got up to defend me. But from the memory of the previous night, I knew he would not succeed.

The lion met Nick at the verandah. There was a terrible fight. Nick was killed. All the time Mama and her friends kept turning the pages of their fashion magazines, talking to each other. I stretched out my hand from underneath the table and Mama took it in an absent minded fashion, without interrupting the conversation with her friends.

The lion moved through the garden room, the dining room, the hall, and then the door opened and he came into the room. At this point I always hoped against hope that the tablecloth might hide me from his sight. But he came straight towards the table. And then he opened his enormous mouth and swallowed me with my feet first. Just when he was about to bite me into two pieces at my waist, I woke up.

The dream really frightened me. Mostly because I knew that there was no escape from it. As soon as I fell asleep it would come, exactly in the same way. It was something that could not be altered; I simply had to go through with it night after night. The fact that the lion did not actually kill me, cut me into two pieces, was no comfort. The horror lay in the slow ritualistic build up, and the terrible feeling that came when the lion swallowed my feet and I knew now would come the moment when he would bite, and I would be dead.

Just as my first dream had given me an insight into madness, so that dream seemed to give me an insight into death, being murdered, killed, destroyed, eliminated, split into two. My ploys of trying to avoid going to bed, just being unavailable when the time came, grew more and more frantic and the grown-ups became more and more exasperated with me. But I never told anybody. There did not seem any point. I do not know why or how the dream suddenly stopped as suddenly as it had started. I only know that by autumn, by the time I started going to school, it had simply disappeared of its own accord.

Jesuits and Nazis

Ever since I can remember, I knew that I would not go to an ordinary school but to the Convent of the English Ladies. Somehow, I don't know why, I resented this. I would, of course, never have spoken about it since it would only have resulted in prolonged accusations about lack of gratitude.

The Convent of the English Ladies goes back to Tudor times, to Mary Ward, who went to St Omer to establish an institution where English aristocrats could send their daughters to receive a Catholic education. Mary Ward's aim had been female education, this meant that her nuns had to mix meditations and prayers with teaching. At the time of her death there were already twenty Houses in Europe. In 1706, the Emperor Joseph II gave permission for a House to be established in Austria and six ladies and two lay sisters came from their motherhouse in Munich.

The Convent was located in a beautiful old baroque building in the middle of the town. I well remember the first time I was taken there. During the whole of the previous year, Grandmother had missed no opportunity to tell me, 'Just wait, your school days will be the best time of your life.' This announcement had made me feel uneasy. I was used to treat such statements made by grown-ups with a certain amount of mistrust. My seclusion from life outside our house, and with it from the company of other children, helped in a way. There were simply no terms of reference. Nobody had ever really spoken to me about the school, nobody who had actually been there as a pupil.

My name had been entered long before we left for our holiday that summer. On the appointed day, Mama and I went to meet the Mother Superior. I had walked past the entrance of the Convent before, seen the little portico, with the dark heavy oak door, that was firmly closed at all times. There was a brightly polished doorbell handle on the left side of the door and when Mama rang, one could hear the sound echoing inside. A Working Sister, dressed in blue, opened the door. We were ushered along a broad corridor, up a dark winding staircase into a room. In the middle, there was a circular seat covered with dark red velvet around a palm tree. We were told to sit down and wait. You always waited when you came to see the Mother Superior. There was an element of well-regulated suspense integral to the life of the Convent and its rituals.

Mama and the Mother Superior exchanged information in dark whispers. Most of it went right over my head. But I remember the end of the conversation. It consisted mainly of questions the Mother Superior addressed to me. She pointed out the painting of Mary Ward that hung on one of the walls. Then she brought out a picture of a little girl with delicate features and golden curls, the kind of little girl I could never be, and told me her story. How she had endured, even courted a great deal of (what I thought wholly unnecessary) hardship to achieve something that seemed to me equally unnecessary and pointless - holiness. I listened politely.

When the Mother Superior had finished her little tale, she gave the picture to me and said, with a gentle smile: 'I am quite sure you too want to become such a good little girl.' I felt it was high time to put the record straight. I answered, with perfect honesty: 'No, Mother Superior, I don't want to become a good little girl, I want to become an intelligent little girl.'

In years to come, Mama used to repeat this story, not without a certain pride in my precociousness, but ostensibly to show what kind of troublesome and embarrassing child I had been. Still, the Mother Superior rose splendidly to the occasion. 'Very well,' she said, without any change of expression. 'If that is what you want, you will have to work hard. But we shall help you.' And with this the audience was terminated.

When we came home, I sat down quietly and listened to Mama giving Grandmother a detailed account of the visit and of my unpredictable behaviour. Then, to my utter amazement, they both looked at me and Mama whispered: 'She looks quite changed, this is the first time something has intimidated that child!' I was speechless. It had not even occurred to me to feel intimidated, or indeed any different, but something in my face and my behaviour had obviously given them the idea. Had perhaps given away something I did not know myself. I felt a new kind of bewilderment, I felt chastised, just a shade humiliated. A suspicion that school was really something quite different. That, for some reason, life would never be the same again. That I had passed an invisible barrier, entered something final, something breaking the existence of childhood, something from which I would have to move, go on, all in one direction. I would never be able to stop it, never be able to go back. It was the end of the tapestry, the beginning of the story. The simple wholeness of the world was forever past.

Despite my isolation for the previous six years, school did not prove traumatic. I have only very vague recollections of my first day. Lots of little girls, a few crying, everybody eventually settling down behind small, time-worn desks that bore signs and symbols carved into them, relics of children who had sat there before. The Convent with its dark passages, unexpected courtyards, the chapel gently glittering in subdued baroque. The chapel could only be reached through a special passage that at one point divided into another, sealed off by a glass door: forbidden to anybody except the nuns. The nuns in their long black dresses, their veils moving noiselessly and with extreme grace - it all fascinated me. Especially our own teacher, young, and (I thought) extremely beautiful, held me spellbound. She had a special habit of dropping her head to one side so that the long black veil fell forward and then moving it back in one graceful gesture. The way she made the cross, not with little crosses on the forehead, the mouth and over the heart, but with one elegant movement touching the forehead and then both her shoulders, was something I began to practise in front of the mirror. I began to dress myself at home in an old black dress that had once belonged to my Grandmother, and with a rosary wound round my waist, I would kneel in a dramatic posture in front of my little red table. Questions about what I was doing I countered with the announcement that I was praying. Praying was a new word, accompanied by a new ritual, and it began to dominate my imagination. Before Christmas, I announced that I would join the Covent and become a nun like our teacher. At home the announcement was greeted just with amusement. 'She'll grow up!' and 'She'll grow out of it!' was all they said. 'It is just a game.' But in the Convent they took me, to my extreme gratification (a gratification mixed with a certain amount of unease) completely seriously.

I think for the first time since I had arrived so unbidden on the scene, Mama was pleased with me. Mater (the Latin for mother as we called our teachers) had told her I showed much promise and was at least two years ahead of everybody else in the class. Mama's new interest and pride in me, had an unpleasant backlash, she began to supervise my homework. Not only did I have to do the section set, but, in addition, one in advance. It was not just enough to fill the page with (at that time, still) beautiful script, I had, in addition, to draw special designs at the top and the bottom of each page. Actually, I did not mind, I found schoolwork quite interesting but the wholly unnecessary pressure exercised by Mama began to irritate me. Sometimes homework ended with shouts and on occasions a slap across the face.

We were at school only in the morning. After lunch I was closeted with Mama in my room and not allowed to leave before everything was done to her satisfaction. Papa tried a few times to intervene on my behalf, saying, I should have some time off in the garden and do the rest of my homework afterwards. But Mama made it quite clear that homework was her domain and she would not tolerate any interference. Halfway through the year she realized, to her horror, that I had absolutely no ear for music; I simply could not reproduce a tune, however hard I tried. I just did not understand how people could distinguish what seemed minute differences between two sounds, and since singing was one of the subjects in which I would be tested at the end of the year, this might seriously endanger my first-in-everything. She actually made herself quite ill trying to make me sing, accusing me of being difficult on purpose. It was just not possible that anybody could sing so much out of tune. 'You can remember a poem after you have heard it only twice,' she would say. 'It is just not possible that you cannot remember a simple tune, you are just lazy.' Eventually I became uneasy too. There was apparently a great deficiency inside me, but I simply did not know what to do about it.

There were about twelve little girls in my class. I noticed them more as a group than as individuals. I had not been used to spending time with children and now that there were only children around, I simply did not know what to talk to them about. Used to the conversation of grown-ups, their talk seemed inept and silly. There was one red-haired girl who took me aside and asked me whether I knew where babies came from. I had never really thought about it, but I did not want to be outwitted. I remembered the incident with the bitch and her puppies, so I just tried to bluff my way out and said, of course, I did, they came from their mother's stomach. 'Ah, yes,' said the girl, 'but do you know how they get there?' I did not of course, so I just shrugged my shoulders and said did it really matter. 'Don't be silly,' said the girl triumphantly, 'I know how they get there!' and then proceeded to give me a detailed (and as I only understood much later) perfectly correct account of the proceedings necessary to accomplish this. I had a clear feeling that she had only made up the whole story to impress me (how could anybody want to do something so stupid?), so when she had finished I just shrugged my shoulders and said: 'Oh that, I knew that long ago!' and ran off.

One girl in our class was the daughter of one of Mama's school friends. I was of course not allowed to go to school alone. Mama would take me there in the morning and fetch me at lunchtime. I had strict instructions not to move from the Convent gate if she was not there in time. Since Mama and her friend wanted to go for a walk and talk to each other, they decided we should be friends. I am sure there was absolutely nothing wrong with the girl (her name was, I remember, Erna) but the force applied to make us friends caused me to revolt and I simply could not stand her. I don't think she liked me either. In any case, day after day, there were Mama and her friend walking behind us, Erna and I in front, never saying a single word to each other. Mama would call out from time to time 'Why don't you talk?' and 'What's the matter, have you lost your voice?' The more she urged me, the more impossible it became for me to address even a single word to Erna, who, no doubt, found me equally loathsome.

I learned the hard way that the other girls could not be trusted. Any remark one made was immediately reported to Mater. I still remember the first incident. Mater had drawn a large E on the blackboard. Without any ill will, I informed the girl sitting next to me that it looked like a hayfork. Immediately her hand went up and she announced: 'Mater, Albertine says your E looks like a hayfork.' I went cold with shock. Even more so when instead of telling her firmly that one did not inform on people (something Grandfather had taught me), my beloved Mater chided me for being rude and unkind. It took me some time to learn my lesson, and to understand that people could take offence about something that was perfectly true and about which one had remarked simply as a matter of fact, without any ill will. Flattery had not been practised in our house and I did not recognize it, even when I realized that teachers and most grown-ups in general seemed to like pleasant variations of the truth better than the simple truth. I grew completely tongue tied when the occasion demanded this kind of deception. I thought, but don't they want to know? What good is the answer if it is not correct?

On the other hand it was just around this time that I told an elaborate and totally untrue story at home and went to a great deal of deception not to be exposed. I have often wondered about this incident, it was more or less a question of unmotivated crime, because I gained absolutely nothing from it. Mama just asked on the way home 'Has anything special happened?' not expecting a positive answer. And then I suddenly realized that I was telling her an elaborate story about the red-haired girl with whom, after the initial discussion, I had never exchanged another word. I said she had 'answered back' to Mater in a frightful fashion, and, getting inspired by my own imagination, I transformed it into a full-scale argument. I did not really expect Mama to believe me but she went home and told Grandmother, and they both were horrified and said this particular girl had always struck them as a 'street girl', it was amazing the Convent had accepted her as a pupil. Now she would surely be expelled. I was beginning to feel highly uncomfortable. Not for a moment had I thought my story would be taken so seriously. Both Mama and Grandmother questioned me about what had actually happened. I felt I had gone too far to turn back and stuck to my original story. I spent a sleepless night. Next day Mama was sure to discuss it with Erna's mother on their walk after school. And Erna's mother would ask the unfortunate Erna who would, no doubt, gleefully expose me as a liar. It was not really so much the thought of being exposed as a liar that worried me. But my lie had been so elaborate and so pointless that Mama would probably never believe I had just invented it, and start endless cross-examinations as to the true nature of what I had told. And if anybody told the girl, her parent might complain and before I knew where I was, I would be expelled.

I felt the only thing to do was to be ill next day. Then Mama would not go to school and she would not meet Erna's mother. In the morning I managed to convince Mama I had one of my sick headaches and was ordered to stay in bed. I pleaded with the Lord in Heaven (the Convent had by then turned me into a believer) that if He let me off this time I would never again tell a lie. The following day I planned to prevent Mama from having her walk with Erna's mother, where the subject would no doubt be discussed. Her interest in the story might die. But when I came out of the Convent I saw the two of them already talking to each other. Since I was not immediately taken to task I thought perhaps she had not found time to discuss the story and muttered something about a toothache. After accusing me that it was probably my fault (had I secretly eaten sweets?) Mama decided to take me to the dentist. The dentist examined me but could find nothing wrong in the place I had indicated.

And then, over lunch, Mama said to Grandmother: 'I asked Erna's mother about the red-haired girl and she said Erna had already mentioned her behaviour several times...' I could hardly believe my ears. I had invented the whole story but the way Mama had told it to her friend had obviously created an association with something quite different. So there would be no further questions, no further inquiries. I had miraculously been spared. I kept my promise to the Lord (at least for the time being), I restrained from further experiments with the truth, I never again told a story for just creating a sensation. But I felt slightly eerie, as if something unnatural had happened. A feeling that increased greatly when, towards the second part of the school year, the red-haired girl was quite suddenly expelled. No reasons were given but I had an uncomfortable feeling that somehow I had been involved. I just could not think how.

Though I had taken to religion in a big way, insisting on going to church not only on Sunday but every morning in the Convent chapel (much to Mama's annoyance since she had to take me school an hour earlier), I had not quite acquired faith. Our lessons in religion were given to us by a Jesuit. (Mary Ward had originally hoped to bring up her girls like Jesuits - something the Vatican had not allowed. As a compromise she insisted that at least all pupils would learn Latin, to make them equal to men. Many years later, when I worked on a critical edition of a 17th century German travel text, I was to be grateful to her.). We had a vividly illustrated catechism and I went along with great enthusiasm studying the stories about Genesis, Adam, Eve and the apple (though I had certain sympathies for the snake, if for no other reason than because nobody else liked it). A serious crisis came when we reached the story of the Flood. The illustration to the text showed a stormy sea, water being whipped around a tiny island where women and children had taken refuge, praying, wringing their hands or cowering on the ground. I took this in my stride until I saw, to my horror, an Alsatian dog trying, obviously unsuccessfully, to crawl up on to the island. The Alsatian looked exactly like Nick.

I bombarded the unfortunate Pater with questions. How could the

Lord in Heaven, in his great mercy, let an Alsatian drown? Surely, the Alsatian had not committed any sin? What sins could an Alsatian commit anyway? The people might have been wicked, but how could an Alsatian be wicked? Surely he could not even suffer from the stigma of original sin? A pained Pater agreed that, no, that was highly unlikely. So why was he being drowned? Pater tried to distract my attention by pointing out the dreadful suffering of the people in the picture. There were women and children, children just like myself, and they were all going to be drowned. I was not remotely interested in women and children. After all, he had so dramatically demonstrated to us that they had all committed great sins and were only getting their just deserts. But what sins had the Alsatian committed? And if he had not committed any sins, how could God, who was all-knowing and all-merciful, let him be drowned alongside all those wicked people? The discussion went on for several days and I don't remember what method Pater eventually used to distract me. (Many years later, when I was already at University, I met him again and he confessed that I had been a horrid little girl and that he could cheerfully have strangled me alongside that Alsatian and whosoever had painted that stupid picture in our catechism. I had, after all, been perfectly right, there was no satisfactory explanation for drowning an Alsatian dog.)

Whereas this had just been a clash of opinions, the second incident involved a clash of faith. Before Christmas we were given a picture of the Holy Virgin and her Child, painted in beautiful lapis lazuli, all surrounded by empty squares. We were encouraged to purchase little golden stamps from our pocket money and put them into the squares so that eventually a golden frame would surround the whole picture. The money we gave would be used to 'save the poor little heathen children'. At first the idea of creating a golden frame around the deep blue picture of Virgin and Child appealed to me. But then I suddenly had real doubts. 'Perhaps the little heathen children don't want to be saved,' I announced. 'Perhaps they are quite happy being heathen children.' The other children gasped. It was not customary in our school to express opinions unasked. My beloved Mater gave me a frosty smile and said I should always try to think before speaking. 'After all,' she continued, 'you say you want to become a nun when you grow up. Perhaps you will become a missionary and go out to Africa and save little heathen children.' 'No,' I protested, 'I don't want to become that kind of nun. I want to become a nun like you.' She just stroked my hair and smiled and said, 'We shall see, the Lord will decide,' and, to my embarrassment, I realized that she thought I was trying to flatter her. I wanted to clarify the situation, explain that I had only told the truth but I realized that it was too late. Whatever I said would only make it worse. I could feel the contemptuous looks of the other girls; I was doing the one thing I most despised in others. At least it looked as if I was doing it, I was currying favours. It took me several weeks to come to terms with it. I would wake up in the middle of the night and suddenly remember it, or it would hit me in the middle of something I was doing. Most of all I felt hurt that Mater could misjudge me to such an extent. Surely if you liked people, you liked them the way they were. Flattery was not necessary between people who liked each other.

Lent came and we began to prepare ourselves for our first Communion. This was to take place right after Easter. A week before we were to go to our first Confession and soon preparations started. We were already well versed in the reality of Hell, Heaven and Purgatory. The number of years allotted for each sin had been used to teach us counting. Taking a sweet without permission warranted 500 years in purgatory, not concentrating properly during mass 1000 years, telling a lie 1500, disobedience towards a parent 2000. Even measured against eternity (which I realized was a very long time) those sentences seemed rather excessive. Especially since something as difficult as refusing to eat a piece of cake on Sunday reduced the sentence by only 500 years. As the numbers were written on the blackboard, the line drawn, the deductions made, I could not help feeling that whatever one did, one was bound to lose in the end. God had a habit of winning that seemed wholly unjustified and unfair. So far we had only dealt with sins such as lying, stealing, not going to mass on Sundays, which corresponded with being disrespectful towards one parents - but now all the other sins were explained to us, first by Pater and then by Mater. Eventually we were given a sheet of paper with all the ten sins written on it. We were told by Mater that we should simply make a cross against the sins we had committed; she would then collect the papers. This was an exercise for the all-important first Confession. Little did we know that we were actually cheated out of one of our basic rights - that of the absolute secrecy of the Confessional.

I could not see what all the fuss was about but the other girls seemed to take it very seriously. There were tears and long whispered consultations with Mater. I had an uncanny feeling that the sixth and the ninth commandment bore relations to what the red-haired girl had told me, but I was not unduly worried. The sheets were collected. Next day there was a fluster of confusion and more whispering. A girl was crying and she was asked to see the Mother Superior. Her mother too was called to the Convent and it eventually transpired that she had put a cross against the sixth commandment. An investigation proved that she had actually thought it meant stealing sweets, and everything returned to normal. The great inquisition closed its books. But the drama was not completely over.

On the very Saturday when we were supposed to go to our first Confession, there was a charity show in the local theatre. A group of little boys and girls, dressed in old Viennese costumes, were called to perform a dance. I was one of the girls. I had no idea how or why I had been chosen. I was not a pretty child and I was not particularly graceful. In fact, I still have a photograph showing a little girl with solemn eyes in a crinoline, a large hat, on the arm of an equally solemn and unprepossessing boy in frock and top hat. The photographer, or Mama, or both, had obviously tried to instil a touch of elegance by making me pick up the hem of my long dress with my fingers. We both looked rather like the wooden dolls one sees in shop windows. But in the Convent, the news caused consternation. Mama was asked to come to the Convent and the Mother Superior invited us both to her audience chamber. She informed us kindly, but firmly, that I would have to go to my first Confession alone, the following Monday. It was unseemly that I should perform so important a ritual on the same day when in the evening I would meet boys. I felt rather important and smug. The only thing that worried me was that I could not understand what exactly I had done to warrant this extra attention. To meet boys? The same evening we had a rehearsal and I looked at the little boy who was my partner, and who had not said a single word to me, nor I to him. For the life of me, I just could not imagine what was so special about him or any of the other boys.

On Friday Mater gave me a little picture of a saint and told me, kindly, not to be too unhappy. This added to my bewilderment. It had not even occurred to me to feel unhappy, just the opposite. I felt rather important.

The charity show and our dance routine went by without anything happening that could explain the mystery. On Monday Mater accompanied me to the chapel. We knelt in an empty pew until the time had come for me to go alone into the dark recess of the Confession room. The Confession was rather an anticlimax. It was taken by Pater, whom I knew so well, he seemed strange behind the grill but not sufficiently strange to turn the incident into an adventure. I faithfully related all my sins, assured him that I had told everything, promised to repent properly and was finally dispatched back into the chapel to kneel beside Mater, who had waited for me, and say the prayers I had been ordered to say. I tried very hard to concentrate because Pater had made it perfectly clear that the Lord would only endorse his absolution if I repented properly. But my mind kept on wandering. Obviously I had done something rather improper and dangerous by meeting 'boys'. So strange that the whole process of the first Confession had been repeated especially for my benefit. Yet Pater had never even mentioned it. I simply could not make out what it was. Something had obviously happened and I had failed to notice it, missed it altogether. Overwhelmed by curiosity, I told Grandfather in the evening and asked him what he thought it was. But he just sighed and said: 'Oh, forget it. Nuns lead strange lives, they are all the time locked up and hardly meet anybody. If people live like that they begin to imagine things.'

I felt a little better. It seemed that after all, the omission was on the side of the nuns, it was not on my side. I think it was from then that my determination to become a nun began to diminish.

In spring a new pupil joined our class. Her name was Irma and she came from Hungary. Her parents had for some reasons sent her to our school to be educated. She was a slim and graceful little girl, with dark hair, shy and she spoke German only haltingly. She was also a boarder. The Convent did not usually accept boarders, except in cases where the parents did not live in the neighbourhood of the town, or could not make other arrangements.

For the first time I felt seriously interested in somebody of my own age. There was a good deal of mystery about Irma. Her insufficient German, the fact that she was a foreigner. Whenever we left school to go home, she disappeared behind the corridor in the Convent, where only the nuns were allowed to go. I began to make a definite effort to talk to her, spend time between classes with her, asked her about Hungary. I was extremely pleased when she responded. I felt she was not somebody who would talk easily about herself but eventually I got to know that she had a brother who was ten years older, that her parents lived on a big estate somewhere in a flat country, and that there were many horses. She shared my interest in animals. Eventually we began to compare notes about Aunt Maria's farm and the thoroughbred horses her parents owned, which seemed so much more fascinating. We spent most of our meagre free time together. I asked Mama whether I could invite her to our home but my request was brushed aside. I was not unduly upset. We had plenty of time. Irma's parents wanted her to be educated in Austria and spend the first four years in the Convent school before attempting the examination for the 'Gymnasium' - just as I was supposed to do. The 'Gymnasium' ended with the 'Matura' when we were eighteen. Time stretched endlessly in front of us. We had all the time in the world. Time to get to know each other, time to become friends. Time to persuade Mama to invite Irma, perhaps even spend a weekend with us.

Towards the end of the school year it was suddenly discovered that I had difficulties with spelling. Mama was horrified. First singing and now spelling. Besides, only stupid children were unable to spell and everybody kept insisting that I was well above average. She would accuse me of being purposely awkward. What made it worse was that sometimes

I would spell the same word correctly, sometimes incorrectly and if asked independently of the text how this word was spelled, I would nearly always give the right answer. Yet whenever a text was put in front of me and I was asked to say whether I had spelled anything incorrectly, I simply did not see my mistake. 'You are too hasty,' Mater would say. 'You must think more carefully, more slowly.' Though she did not actually accuse me of doing it on purpose, I could feel that she too thought I was not taking enough trouble. For the first time in my life I was feeling confused. I did not understand what all the fuss was about. Surely if I wrote *un* instead of *und*, everybody could see I had simply forgotten the *d*. The meaning of the sentence was perfectly clear. Why all the fuss about an oversight that made no difference at all. The completeness of the thought was surely more important than a small detail.

Mama began to give me endless dictations each afternoon. But just after I had produced a page of faultless text and we would both relax, I would misspell a word I had already written down correctly several times before. What infuriated her most was that sometimes I would simply misplace the order of the letters in a word. 'This is ridiculous,' she would say. 'You can't tell me you cannot see what you are doing. You just want to annoy me.' And Mater would insist: 'Try harder, you must try harder.' I crawled angrily back into myself and into my friendship with Irma.

Irma's spelling was much worse than mine. But her mother tongue was Hungarian. Still, everybody used to make fun of her most glaring mistakes. I found this exceedingly mean. My beloved Mater did not intervene openly, tell them not to laugh about Irma and slowly but surely my admiration for her began to wilt. She stopped being my role model. Somehow I felt disloyal about it and tried even harder to please her. But it was no longer the same. The magic thread had been stretched too far. As long as I had felt total admiration for her, I had felt no need to please her. It was only now, when those feelings had gone, that I found myself performing empty gestures. This bewildered me even further. I had my first suspicion that truth was not a simple 'yes' or 'no' but a variation of many factors.

My spirit rose when I learned that I was to have a part in the end-of-the-year play. The Convent had its own theatre with a proper little stage. The play was about a mixed marriage. A Catholic girl had, against the advice of all who knew better, married a Protestant. The grown-ups were played by the girls from the 'Gymnasium' who were in their last year and already eighteen. As far as I was concerned they did, of course, not play grown-ups, they were grown-ups. My part was that of the little daughter. I don't remember how the play ended. Since staying married to a Protestant, or being divorced, were equally deadly sins, I could hardly imagine a satisfactory conclusion. I remember there was one scene where the wife bursts crying into her daughter's room, hiding the picture of the Virgin Mary in the little girl's bed, because her husband had strictly forbidden her to keep it in the house. There was even a part where I had to sing (together with the girl who played the wife). Mama was mortified. How anybody could have given me a part where I had to sing was beyond her imagination. I was bound to look ridiculous. She was going to sit in the last row, leave at the time in question, anything rather than live through the humiliation of having people make fun of her child. She made me practise the song again and again. This did nothing to ease her worries because I did, of course, sing horribly out of tune for most of the time.

On the evening of the performance we were made-up in the dressing room by a professional dresser, who had come over from the town's theatre. She shook her head at the way my hair was pulled back into two thick plaits and went into the older girls' room to borrow a curling iron. Without much interest I let her make-up my face, comb my hair and arrange it into loose curls round my shoulders. 'There,' she said. 'Take a look at yourself, that's much better.' I looked into the mirror. The girl who looked back at me was strange, yet familiar. Was it really I? And how had the transformation taken place? For the first time in my life I thought of attractiveness in relation to myself. That one could actually do something and simply bring it about. I had a feeling of many doors, many corridors, all only dimly perceived, many possibilities that lay in the future. I felt suddenly happy, intoxicated, eager for the play to start, eager to walk on the stage, even eager to sing my little song. I was quite confident I could do it and, by sheer coincidence (or miracle), I did indeed manage it.

Mama was delighted. Even more so when next day it turned out that I had got the first place in class - though with the recommendation that I should practise my spelling over the summer holidays.

In this way my first year in school ended. Irma went back to Hungary and we managed to overcome our mutual shyness enough to ensure each other that we were looking forward to meeting again in the autumn. Mama gave me a long lecture that I should not let my head be turned by praise, the fact that I had done so well in the play, and equally well at the examinations, was pure luck. We were now going to do two hours dictation every day. Those two hours became quite disastrous. Mama would get more and more impatient, and the more impatient she became, the more spelling mistakes I made. Fortunately for both of us it was soon time to visit Aunt Maria in the country and this put an end to all dictations for the time being. (I did not know it then but this was our last visit to Aunt Maria. Sometime before Christmas she died, but I never found out how and why. Nobody ever told me about it and I knew it was pointless to ask questions.)

When I returned to the Convent in the autumn, nothing was quite the same. Mater had been sent to Rome and in her place was now a new teacher - not a nun. Normally a teacher would follow her class up to the end of primary education; at least, that was what I had always been told by Mama. It seemed that most of the other mothers were highly indignant but the Mother Superior explained, gently but firmly, that the decision could not be revoked. Mama lamented that the change would do me great harm and expressed her surprise that I took it so calmly. But I was not really greatly perturbed. Somehow, during the later part of the year, my admiration for Mater had diminished. Many small incidents had exposed a nature not quite as perfect as I had thought. She was no longer the shining example on which I wanted to model my life.

The second change was more perturbing. Irma was not there at mass that first morning, nor was she there on any of the following days. It turned out she had drowned during a boating accident. Nobody ever told me the news openly and officially, I just heard it from remarks Mama made, not even addressed to me. This gave me the feeling that her death, and my loss of her, was something I could not even openly acknowledge. It almost had the aura of a guilty secret. It was a truth I had gained by accident and this debarred me from recognizing it. I was bewildered and not quite able to grasp what I really felt about it. It was my first introduction to loss, death, disaster, an introduction to the hurt that can arise if one allows oneself the luxury, or the carelessness, of emotions. I had a suspicion that people were in reality all strangers who came and went at will, that one had no control over their movements. Her death seemed almost like a betrayal. The situation was more confused by the fact that nobody seemed to have taken our friendship in the least seriously. It had not even warranted the consideration of official information. Mama put down my spells of dark mood entirely to the loss of Mater. There was obviously a large margin of misunderstanding within all human relationships; I received sympathy for something I did not feel, complete indifference for what I did feel.

Most of the mothers were openly mistrustful of the new teacher. The others probably more so than Mama. I had not been sent to the Convent because my parents were devoted Catholics, but simply because it was academically the best school available. But many of the others felt their children ought to be taught be a nun, not by a secular teacher. After all, secular teachers would have been available in non-fee paying state schools. I was no longer willing to be so easily impressed by a person, simply because this person introduced me to a world of new and fascinating information. I had tentatively learned to distinguish between the person and the work. But slowly a sort of mutual understanding began to develop between Frau Doctor and myself. Not sentimental, not emotional like last year's infatuation with Mater. I simply began to realize that Frau Doctor was prepared to give me a fair deal. Instead of constantly and gently subjecting me to humiliation about my weakness in spelling, she simply told me, casually, not to worry, many successful people never learned to spell perfectly, if I concentrated on what I could do best, the problem would solve itself. I felt the first, tentative, stirring of self-confidence. Before Christmas she suggested that Mama should buy me a certain book. This book was to exercise a profound influence on me. It was to occupy my mind for years to come, opening the view to an enormous vista, arousing not only my curiosity but also my desire to test information, not just to receive and accept it. There were altogether three volumes. They told the story of two children, a boy and a girl, whom the grandmother of the boy took through a gorge to save them from some never clearly mentioned danger. An avalanche killed the grandmother and the two children were imprisoned in the valley into which the gorge led. There they lived through the whole development of the human race from making stone tools to eventually building a stone house. (It was probably this book that made me choose ethnology at the university of Vienna - and motivated me to undertake my many journeys.)

Mama was still trying to interest me in a friendship with Erna, though with less enthusiasm. There were walks after school and on Sundays after mass, but Erna and I remained obstinately silent. Not exactly on purpose but there was simply nothing I could think of saying to her. She was neither particularly interesting nor obviously intelligent but she had a talent for playing the piano. Her mother hoped she would be able to pass the entrance examination into the 'Gymnasium' and stay at least until she was sixteen. After sixteen she could go to the Music Academy. I did not know how great Erna's talent was, mostly because I still had absolutely no ear for music. But I noticed that Mama countered the plans Erma's mother had for her daughter by suddenly claiming she had always considered a university education for me. I would become a Frau Doctor like our teacher. She had no preference for any particular discipline in which I would gain this distinction but she made it quite clear that the matter was settled.

I cannot really remember any of the other girls who stayed with me in the same class until we were ten. I still had nothing in common with them. Moreover I loathed their squeamishness, their crying when they were in the slightest hurt, or even if somebody just said something they did not like. I had never been given to crying. Grandfather despised weakness and Mama, if as very small child I had cried without what she considered a sufficient reason, would just cuff my ears 'to give me something to cry about'. Not having been reared on sympathy, I had none to give.

I did, however, about this time, acquire if not an actual friend at least a play companion. A new family had moved into the house next door, the daughter was four years older and therefore already impossibly 'old'. She had no interest in little girls and I hardly noticed her. But the son was one year younger than I and he soon became my devoted follower. There was never any question of friendship. Friendship, I already knew, was only possible between equals and Hans was not my equal. In the world in which I had grown up superiority was associated with age and anybody younger than I could not be my equal. He also went to an ordinary school and was therefore (according to the dictum of my upbringing) less intelligent.

Hans accepted without question that I was the leader, another reason against the possibility of friendship. I was, of course, not allowed to go into his garden (the walls of our garden were still a demarcation line I was not allowed to cross without the company of an adult), and Hans was not allowed to come into our garden. But we began to talk to each other through the fence. Eventually we would sit on the wall in the back of the garden where a large tree (we called it the 'paradise tree' because of its beautiful blossoms) shielded us from view. I decided what games we would play, usually war games. From my cave children book I had learned how to fashion a bow and arrows. Hans, who had the liberty of his father's garage, would bring along tools and use them according to my instructions. He never questioned my authority or the fact that it was up to me to decide whether I wanted to play with him or not. Eventually the roof of their garage became another no-mans land from where we could shoot our arrows. From time to time we tried to play in our garden but this was usually stopped by Mama. She felt that Hans was too stupid for me to play with, or by Grandmother who felt that I should play with little girls (if at all) but not with little boys. 'I don't like little girls,' I would protest. 'They do nothing but cry. At least Hans never cries.' This, together with the fact that, for the first time in my life, I had somebody to whom I could give orders and who followed them without question, was the main reason why I accepted Hans as a play companion in the first place.

In winter we would pelt each other with snowballs or build fortresses from frozen snow. We also had some other, more complicated games. I remember once, after I had given Hans detailed instructions about a campaign we were fighting, I finished my sentence by shouting at him over the fence: 'And don't forget you are my enemy and I am your enemy.' Grandfather who was working in the garden at the time rebuked me mildly and pointed out that Hans was definitely not my enemy. 'But we are only playing,' I cried, exasperated at the incapacity of grown ups ever to understand anything that was perfectly clear and simple. And Grandfather sighed and asked whether I could not think up some friendlier games. 'But you were in the war,' I said. 'You told me about it.' And he sighed again and said something about wondering whether that had not been a mistake. He also said something about sooner or later there might be a real war. 'But then it's a good thing you told me,' I assured him. 'I shall know what to do.'

There was talk about war. There was also talk about what the grown-ups called 'troubles.' Troubles included anything from the broken glass windows of certain shops one saw sometimes in the morning, or swastikas painted on fences and garden walls. Once I was woken up by the sound of shooting in our garden. Papa came and took me into their bedroom and told Mama and me to lie on the floor, he would see what was going on. But when he went to the window I suddenly cried: 'Don't go Papa, they'll shoot you!' and was surprised by the horror I suddenly felt. With the danger of death from the dark garden, Papa's life became suddenly more important to me than it had ever been.

Grandfather seemed to become more withdrawn, almost as if something was weighing him down. We no longer went for our walks but he began to teach me chess instead. Every evening I would spend an hour in his study over the chessboard. The figures had a special history. Grandfather had carved them during his time in the prison camp in Siberia and then brought them with him on the long journey home. But he no longer told me about the camp. It was as if the word itself brought on a frozen look and I stopped asking questions. I liked our hour of chess. Of course, I never won but then I did not expect to. It gave me a feeling of being allowed a short visit into the world of adults, not senseless adults like Mama and Grandmother or Erna's mother. But adults like Grandfather, Frau Doctor, people who mattered. I was aware of serving an apprenticeship - I just did not know for what.

I do not think that even at this stage I was still a true believer. I tried to believe on a conscious level but I knew that deep down I never quite succeeded. Even after the Confessional I was spending most of my time wondering whether I was truly repenting and whether there would be true forgiveness. But the Confessional never held more fear than an examination. Pater was simply my teacher and I was more nervous whether I was giving the expected answers than whether I was telling the truth. I simply did not know what the truth was and I was not really capable of proper repentance. From Mama I had learned that punishment followed the crime and that repentance could never avoid punishment, so what was the point of repentance? I transferred the absolute authority of Mama to the absolute authority of God and I had a suspicion that in the end he would use his powers in the same way. The figure of Christ held no fascination; suffering was in my mind already associated with weakness. I felt a certain interest in the figure of the devil, for being the only one who had made an attempt at independence (but, if God was really all-powerful and all-knowing, how could he have created the devil in the first place who, he knew, would rebel, and how could he be all-merciful and create somebody who to his knowledge would have to suffer in hell for all eternity?) Of the two the devil seemed more interesting and more fascinating. For the first time I was fascinated by evil. Not because I wanted to commit evil but because I wanted to know about it.

Since there were no emotional ties between the teacher and myself I felt more comfortable than during my first year at school. I did not really feel the need for friendship either. My mind was filled with the possibilities of new discoveries. The three volumes about the cave children that portrayed the whole development of the human race from early toolmakers, kept me fascinated. I began to experiment, test the truth of the written text. I spent hours drilling a hole into a piece of stone with bow and arrow, the way it was described in the first book, I shaped stone tools, made baskets from dried straw, fashioned a small weaving frame. My favourite place became the verandah. The tools I had made were kept in the little cellar underneath it, the same, where, when younger, I used to hide in order to avoid the afternoon siesta. The cellar had the appearance of a cave, the cave described as the first living place in the beginning of the book. The grown-ups did not interfere; I was out of the way and quiet. Only Grandmother would protest occasionally that it was just not right that Hans should come over the fence from time to time and that we should both disappear in the cellar. It was not nice, she insisted, for a boy and a girl to go into a cellar. This remark seemed so senseless and idiotic to me that I paid no attention. Grandmother was beginning to get more eccentric, she was putting on a lot of weight, something she tried to ignore and any reference to it was considered sheer impertinence if not altogether a lie.

I lived in an inner world that was rapidly expanding, that included increasing amounts of new information, new vistas, the things I learned at school, my experiments trying to test the incidents described in the cave children's books, the chess evenings with Grandfather. The grown-ups too seemed to become more preoccupied and less interested in what I was doing. Grandfather's godson (I began to doubt that he was actually his godson), the one who wore white knee-socks and cornflowers, used to come more often, and the two would closet themselves in Grandfather's study for hours. I realize now that they were truly fond of each other. Francis was becoming more and more involved with the illegal Nazi Party. He was trying to convince Grandfather that there was a lot of good in it, that Austria would have a brighter future as part of Germany, that the stories about persecution in Germany were nothing but propaganda. Grandfather just listened and seemed to grow more quiet and sad but he never tried to convince Francis that he was wrong. Grandfather never tried to impose his opinion on others; he had absolute faith in people's reasoning power and felt that they would eventually find out for themselves.

I knew about all this vaguely but I had, of course, no idea of its implications. It was the habit of grown-ups to worry about things that seemed meaningless. I did not yet have a clear concept of the future; as a matter of fact I hardly knew what future meant. But it is possible that during those few years I noticed more, and forgot more, than I remember, because there seems a curious amnesia, a curious blank space whenever I try to think of this period. I remember hardly any incidents. In fact I remember less than I remember about the previous years, when my powers of understanding and observation must have been much less developed. I still did not know what Papa did, I only knew that he seemed to be away from home most of the time.

And then the first cracks appeared in my childhood. The events of history with their relentless continuity forced themselves into the magic circle, where all things existed simultaneously and everything was already mine because it only existed in my imagination. Though when it happened I did of course not understand its significance. We never do. The important events, the great decisions, begin with everyday little incidents, like a forgotten ticket, a lost handkerchief. They are so small that we cannot possibly recognize them as the riders heralding the coming of great changes, and since they seem so insignificant we cannot guard ourselves against them. So when the so-called great decision finally comes, it is much too late, because the actual decision had been taken long ago, at a moment when we knew nothing about its consequences.

I remember that particular day quite clearly, in one logical sequence from morning to evening. Our school had uniforms but we wore them only on special occasions: a blue pleated skirt and a white blouse with a sailor's collar. The idea of a school uniform was no doubt connected with the English school from which our house took its origin. Why it should be worn only on special occasions was something I never quite understood. When I asked Mama it became clear that she did not know either. That morning she helped me put it on and then we left to go to mass before a special occasion. What the special occasion was I have long forgotten.

When I think of that day I remember it in colours: the dark navy blue of my skirt, the greyness of the sky. Grey seemed the predominant colour, a background against which all other colours stood out in a single event. When we passed the church of St. Joseph which had been finished the year I was born, and where my baptism was registered as the very first that ever took place there (a fact which always filled me with a certain satisfaction), we found the glass windows broken and a swastika drawn in black on the church door. As we went further we found a blood red flag with a white circle and another black swastika in the middle, hanging from the branches of a chestnut tree in the avenue that lined part of my way to school.

Mama murmured: 'Oh God, they will go on like this until there is real trouble.' Just as she said it, we saw a man across the road, he wore a brown shirt and an armband with another swastika. I pointed at him but before I could say anything, Mama hissed: 'Just come along and don't look at him.' We continued our walk but the greyness of the sky slowly mingled with a suspicion of disaster. What disaster I did not know but I could feel it clearly and so, I think, did Mama. Only just a quarter to seven but there were more cars, more people, and more movement. And there was a strange kind of excitement, a subdued excitement that seemed beyond clear expression. A woman, she was fat and ugly, came up to us and said triumphantly to Mama: 'You'd better go home. There won't be any Convent school today.' Mama's face grew stony; she grabbed my hand and walked past the woman, utterly ignoring her. 'Well, see for yourself,' shouted the woman. 'It's finished with all those damned privileges. You'll have to send your brat to an ordinary school now,' and when Mama still did not yield, she shouted, 'Hitler is here, don't you know?' Mama's grasp of my hand became firmer and we continued to walk on. I was not really worried. I was used to grown-ups behaving in a wholly illogical fashion from time to time. We marched on. There were more people. More men in brown shirts. It seemed to me that some of them viewed us with a certain amusement but I followed Mama's example and ignored them all. Thus we suddenly came face to face with one of Grandfather's friends. He looked at us horror struck. 'What on earth are you doing walking her around in that outfit?' he demanded, pointing at my school uniform. Mama's icy remoteness collapsed. In a pathetic voice she related our encounter with 'that vulgar woman' and asked for an explanation. 'But haven't you listened to the radio?' asked Grandfather's friend in disbelief. Mama admitted that, for one reason or another, nobody had. 'The Germans have marched into Austria this morning,' said Grandfather's friend. 'If I were you I would go home through the park and get your daughter out of that school uniform.' 'Marched into Austria?' Mama cried, and threw her arms around Grandfather's friend. 'If this is true, I don't want to live.' Grandfather's friend pushed her away. 'Don't stand around here talking to me. I am Jewish and I have been told I am already on their black lists.' By the time we reached our street, Grandfather was already pacing up and down outside the garden gate. After we had left, he had listened to the radio and heard the news. I was straight away divested of my uniform and sent to my room. In the evening, Papa came home. The rest I remember is airplanes, airplanes flying endlessly across the sky. I watched them from the balcony in front of my room, hanging on to Nick's fur for comfort. There was much shouting, much talking, much whispering and much quarrelling in our house. For the first time I felt wholly excluded. A few days later Hitler drove through our town in his triumphal procession to Vienna. Papa took me to see him. 'It is a historic occasion', he said, brushing aside Mama's and Grandfather's protests. It was the first time he had asserted himself in that house that was not his own. To my surprise, they all remained silent and let him take me with him.

Hitler's route went straight through our town. There were masses of people waiting and Papa and I joined the crowd at a strategic point in the Wienerstrasse. We had to wait a long time. Eventually the convoy came. Cars, outriders on motorcycles (all in black), and then an open car. In it stood a very ordinary looking man, stock-still, his right hand lifted in an unwavering salute. Papa lifted me up so that I could see it all better and then, as swiftly as it had come, it was all gone again. We went home.

Mama had retired into the bedroom and was obviously in no mood to speak to Papa. Nobody mentioned the incident. Papa and Grandfather had a short conversation and I heard Papa say: 'Just one sharp shooter from any of the attic windows could have killed him.' There was regret in his voice about the lost chance and a certain kind of respect for the courage of the man who had chosen to ignore this possibility.

The upheavals that followed came in such quick succession that I cannot really recall them in their proper order. The nuns were barred from teaching. The part that housed the primary school, the park and the large 'Gymnasium' on the other side were confiscated. The nuns were no longer allowed to enter this part. However, since a secular teacher had, since the departure of Mater to Rome, taught our class, the change did not affect us in any noticeable form. For the time being Frau Doctor remained our teacher. There were no longer lessons set aside for religious instruction. Instead we had more sports lessons and we were introduced to ball games and competitive sport, which I thoroughly detested. The most noticeable change was the influx of new students. The new order was adverse to privileges, we were told, privileges based on money and social background. Privileges belonged to those who deserved them, who were in themselves superior. From 12 our class rose to 40 girls. Most of them came from the surrounding countryside, village schools, and poorer families in town. School fees were abolished,

I could not find anything of interest in the new girls. If anything they seemed less intelligent, less well informed than our original group. But since I had felt no need to make friends before, their arrival held no special significance for me. During the first year the new authorities were busy reorganizing the school system and we were not presented with many new values. The only definite change I remember was that we were no longer 'Austrians' but 'Germans'. The term 'Oesterreich' was strictly forbidden. It was now called 'Ostmark' - the eastern part of the Reich.

At home the situation was more complicated. Politics and religion had so far played little part in our household. Or I should perhaps say that, since they were subjects rarely mentioned, I had assumed that their roles in our lives were more or less negligible. Now I learned differently.

I could not help noticing the broken glass windows in some shops, the people with yellow arm bands, the burnt-out synagogue, and the people who were occasionally paraded standing outside a shop between two black-clad men with a placard on their back saying 'This German pig has bought from a Jew.' Then five girls disappeared from our class. I would probably not have noticed their absence; I had hardly ever spoken to them. But next morning we were informed they had been removed from the school because they were Jewish. This was followed by a homily that it was inadvisable to teach Aryan girls next to Jewish ones. Another girl was pointed out to us as being one quarter Jewish. But, we were told, because of the three quarter good Aryan blood (that outweighed the quarter of bad Jewish blood) she would be allowed to continue with her education as long as she proved herself worthy.

I mentioned the incident at dinner with dramatic effects. Mama started crying in a frantic, hysterical way, accusing Papa that these were the people to whom he had sold himself. Papa protested that he had done this only to protect us all, and Grandmother accusingly said that 'This child is going to bring us all to a bad end.' Mama rushed upstairs. Papa followed. Grandmother kept on complaining that I was, as she had always foreseen, a disaster to the whole family. It was left to Grandfather to take me into his study and explain the situation.

I remember the evening, though I totally failed to understand the significance of it. The subject was never discussed again and after some time I more or less forgot about it. Grandfather said that his mother had been born in Poland, in a small place. The church where she had been baptized no longer existed. In other words there was no proof that she had not been Jewish. I said: 'Was she Jewish?' Grandfather hesitated for a moment and then said: 'No, but we cannot prove it.' I did not quite see the logic. 'But that's all right.' - 'Well, not quite,' said Grandfather. 'You will have to take a certificate of your Aryan descent to school. There is no document available about your great grandmother.' So, he continued, I must never mention this to anybody. Would I promise to do this? Still bewildered, I assured him that this would not be difficult. But why had Mama accused Papa of having sold out? Grandfather agreed there was a second problem. Papa had joined Hitler's Party (he did not say when and I was not terribly interested). Most of our friends, continued Grandfather, were strongly anti-Nazi. So I was not to mention this fact to anybody either.

I did not really understand the logic of these two arguments. But I could not help feeling a certain contempt for the grown-ups who were messing up their lives in the most ridiculous fashion. But I was used to isolation. I had never had the wish to be 'like everybody else'; I had never really wanted to join in the games of the other girls. I had always, though never expressing it to myself in so many words, considered myself 'different'. Now Grandfather supplied the proof that I was right. But I failed to be pleased about it. Somehow I just could not figure out all the connections.

During the following year the uneasy situation in our house increased. We were now divided into so many factions. All claiming, like the characters in O'Neil's *Long Day's Journey into Night* to act for the mutual benefit, all creating fictions by their very actions. In a way I was now more superfluous than ever. I still felt protected by Nick but I also felt a certain unease.

And then the war, about which there had been so much whispering and veiled prophesies, started. I remember the preceding night very clearly. I suddenly woke up, the moonlight fell into my room, and then I remembered what had awoken me. There were women and children crying in our garden. Lamenting. I could hear them clearly and yet there was vagueness about it. There were no individual voices, just a sort of communal lament, a sum of total suffering.

I lay absolutely motionless. I knew that unknown forces were torturing them and sooner or later somebody would come into the house, my room, and do the same to me. I was paralyzed, not so much by fear, but by the sudden reality of events. I had no idea what the reason was for all the suffering, who the people in our garden were, how they had got there in the first place. I just felt that it was not my task to know about it, I was only a child, and this was definitely grown-up business. But I also felt that since I knew about it I had to warn the grown-ups, tell them about it.

I climbed out of bed and went to my parents' room, a feat normally forbidden by sentence of death. 'There are people crying in our garden,' I whispered. I had not dared to put on the light, because I clearly felt that if my warning had to be effective nobody outside the house must know about it. 'They are doing dreadful things to them, something is happening.' Papa switched on the lamp at his bedside table and went to the window. 'But there is nothing in our garden', he told me. 'Come and look.' I recoiled. Strangely he had not lost his temper. He just said: 'Listen, there is not even a sound.' I listened, it was true. The dreadful, communal lament had stopped. I ventured to the window. The garden, so familiar, was bathed in white moonlight. Not a sound, not a shadow. Yet nobody reproached me. I was even allowed to sleep on the divan in my parents' bedroom, a rare treat.

In the morning a telegram arrived for Papa. The war had started in the early morning, at 7 o'clock he received his call up. He left the same evening. (Many years later I was in Tiberius after the Lot airport massacre. We had not seen any newspapers in the evening but the imminent conflict was known to all of us. In the middle of the night I woke up, and then I heard it again: the disembodied yet only too real lament of women and children, right outside the windows of our hotel;)

As soon as Papa had left (not for the front but for some place in Czechoslovakia), Mama's attitude towards him underwent a complete change. She, who had for years refused to leave her parents' home because of him, suddenly professed, loudly and tearfully, her great love, her great suffering on account of his absence. Tearfully she demanded consideration from the rest of the family and all our family friends. Indeed anybody she met was immediately treated to the tale of how Papa had been called up only two hours after the outbreak of war. This was viewed entirely as her great misfortune - not Papa's. All men not yet in the war (and of course at that time many were not) were viewed as a personal affront as far as she was concerned. I was astonished. I had witnessed all the scenes, Papa had often been away before, and it had never in the least bothered her. And now suddenly this complete devotion to somebody she had so far treated, at best, with tolerant contempt. For once she and Grandmother were divided on the subject of Papa. Grandmother found the whole display of grief ridiculous but Mama insisted on her part as the unfortunate wife whose husband was away in the war. Grandmother's hostility towards Papa increased. Partly perhaps because Papa had been posted to a place very near the estate where her mother had once lived, in great poverty.

I remember the following afternoon. I sat on the balcony in front of my room and said to myself: there is a war, like the wars about which we learned in school. I felt a certain excitement about not just reading about history but living it. For the first time in my life I was part of history, later generations would count me as part of it. Some smoke rose at the end of our neighbour's garden. I wondered whether this was already a bomb dropped by the enemy. But then Hans appeared from the bushes and called up to me saying he had started a bonfire, and would I like to join him? Slightly abashed at the banality of what I had considered a historical event, I climbed down the balcony outside (so as not to be seen by the grown-ups), went over the fence and joined him.

There were more changes at school. The place of the nuns was now firmly taken by teachers from Germany. Mostly, it seemed to me, male teachers. Frau Doktor had left and our teacher sometimes made a pupil kneel in front of his desk as a form of punishment; he even slapped some of us occasionally. Austrian law had strictly prohibited any physical punishment but it had not been difficult for the nuns to establish complete authority over us. As a matter of fact we had been more in awe of them (they always seemed to know what we wanted to do half an hour before we even thought about it) than of these rather crude individuals. I decided the safest way was to draw as little attention to myself as possible.

At this time I was given another book that was to have great influence on my life. It was a collection of tales from Germanic mythology. I think we were told at school that out parents should buy it for us, but I am not sure. I was truly fascinated by the various Gods, by the violent Thor, but most of all by the deadly Locki. Baldur bored me, just as hardly a year earlier Christ had bored me, when compared with the more intricate figure of the devil. I preferred the great lords of darkness to the passive sun gods. Just as I had found Hagen more interesting than Siegfried. I began to write poems, devotional poems to the great gods of destruction. It was my first venture into poetry, into the reality of mystic exaltation. (Strangely, when I eventually fell in love for the first time it was with one of those passive sun gods and not with any of the great figures of darkness - or so at least it seemed to me.)

The last year of my junior education was overshadowed by preparations for the entrance examination into the 'Gymnasium'. The 'Gymnasium' too had been taken away from the nuns, all the teachers were secular and the director was somebody from our town. I would have to spend eight years there before passing the 'Matura' to qualify for university. I felt a certain resentment at this plan. First of all because Mama had obviously planned it since my birth, and I had a suspicion that anything Mama planned for me would not really be worth having; but I knew better than to voice such sentiments. Once, when asked by one of Mama's friends to what school I would go, I let slip: 'I shall have to go to the 'Gymnasium'.' I got thoroughly scolded. Mama pointed out how grateful other children would be if they were allowed to go to the

'Gymnasium' and I had said I 'had to go'. This also provided ammunition for Mama's friend who said such an education was in any case wasted on a girl who would, after all, only get married and have children. Here at least Mama and I agreed. Though I felt ambivalent about marriage I was absolutely determined not to have children. Both Grandmother and Mama had imprinted on me that motherhood equalled torture, it was not only unrewarding and degrading, and it practically put a death sentence on you. My dolls had not amused me, the idea that a lifeless figure could become a screaming demanding baby, that wanted to be washed and fed, did not make it more attractive. The facts surrounding birth were still a mystery to me but I knew it was a horrible mystery. I had at one point been convinced the baby would be born through the mouth, you vomited it out. I often lay in bed wondering how anything so big could come through your throat. It would be worse, much worse than those occasional attacks of headache, nausea and vomiting I sometimes experienced. I was determined to avoid it at all costs. Besides, looking at Mama and her perpetual misery I had no wish to become like her. Grandmother too was not exactly an inspiring example. I had no aversion to sex, which I vaguely imagined would be just one short penetration (where I was not quite sure), but pregnancy was to be avoided

I don't think that in those first ten years of my life I had really left the garden and entered the world. I had made no friends. Irma had died before the promise of friendship had become a reality. Hans was unquestionably not an equal, he was just a sort of slave who did my bidding whenever I wanted to play with him. Grandfather was withdrawing into himself. Grandmother and Mama were involved into their own secret war, in which we were all pawns like those Grandfather and I used in our chess games. The only true friend was Nick.

The entrance examination to the 'Gymnasium' made me realize, for the first time, that outside events had their own power. I had lost some of my original confidence. The German teachers, introduced since the occupation of Austria, had not pronounced me ahead of the rest of the class the way Mater and Frau Doctor had done. There were children whose fathers held positions in the Party and they could claim special attention. Sport had become all-important and I not only hated sport, I had no talent for it whatsoever. Even my handwriting was beginning to get worse.

The examination consisted of mathematics, retelling a story and an essay about our last holiday. I was uneasy about my spelling and so was Mama. I still found it difficult to associate certain sounds with written symbols. I knew what Mama's reaction would be if I failed. Much worse than anything I had ever experienced.

About forty girls had come to take the examination, only five were from our original Convent school, none of them was my friend. Erna was there but her mother and Mama had given up the unequal task of making us friends. Part of the examination was in the morning; in the afternoon we had mainly oral tests. Our mothers were there too, the ones from the original Convent school in a corner, conspicuously ignored by the rest. Some girls did not seem in the least intimidated, they chased each other around through the school building, screaming and laughing. There was one girl who looked at least 14 and who seemed to be the ringleader. Mama said to Erna's mother: 'Well, it won't be difficult for her to pass. She must be at least fourteen.' An elderly woman with a hooked nose turned round and said in an offended voice: 'Excuse me, she happens to be my daughter and she is ten just like everybody else.' At this point I was called into the examination room and asked to write several words on the black board. I was terrified. I had heard somebody say that those called were undecided cases. Having executed my task, a short man (it later turned out that he was the director of the 'Gymnasium') sighed and said: 'Well, why didn't you write it like this in the first place. I suppose you were nervous.' I whispered: 'Yes.' And after a short sermon that a German girl was never nervous, I was dismissed.

Mama waited at the exit and began to cross-examine me. I answered, more or less truthfully, that I had only been asked to write some words on the blackboard and that I had done it correctly. She was so relieved she forgot to ask me whether I was telling the truth.

The nightmare dissolved - I was among the ones who had passed. Erna had failed and cried bitterly, she would now not be able to go to the Music Academy. I felt sorry for her and slightly ashamed about all those years when I had refused to make friends with her.

The war had at least one pleasant side effect. Papa was stationed in Czechoslovakia and by the time of the first summer holiday, Mama finally agreed that we would join him. This holiday was a long reprieve from the new realities of life that were crowding in on me. It was the first time Mama and I went alone on a journey. Mama made the best of her newly imposed hardship by telling everybody in our compartment that she was travelling alone because her husband was 'in the war'. So far she had refused to see him but now she suddenly treated the separation as a tragic twist in her life. I had the uneasy feeling that most people in the train were bored by her tale. However, in Linz, where we had to change from the train to a bus, two young officers transported our entire luggage and Mama blushed and behaved in an entirely different manner. I suddenly realized that she felt flattered by their attention. This was a wholly new trend in her character. The journey in the bus was long and hot and to Mama's annoyance I was sick several times on the way.

Those two months in Czechoslovakia introduced me to a different way of life. Papa worked in what I supposed was some kind of headquarters and we stayed in a little flat in the house nearby. I somehow fell in line with the village children and we went swimming in the nearby lakes. There was also Herman the son of - well, I was never quite sure who his father was. But his mother seemed to be a friend of Papa's and, surprisingly, Mama became her friend too. I did not much like Herman. He was four years older than I and he wore glasses. He also had a funny way of talking to me that somehow made me feel uncomfortable.

In the 'Gymnasium' too there were changes though I was at first slow to notice them. In fact it was a relatively small incident that brought them home to me. All girls whose fathers were in the army, and whose performance in school was above average, were to receive a special sum of money at the end of the first term. When my name was first mentioned I felt actually quite pleased with myself. I felt it was a form of distinction. On the appointed day we all assembled in the great hall and when my name was called I walked up to the podium and took the envelope. It was only when I walked back I heard one of the older girls hiss at me: 'Say at least thank you when you receive charity!'

I was absolutely stunned. I had never before received charity. Charity was the sum of money Mama used to hand over to the nuns for poor children. Was I now a poor child? Obviously. I had no idea what had brought about this fall from grace.

On the way home I brooded over the remark, which I knew, was meant as a rebuff. Slowly I began to understand certain things in retrospect. Obviously the reason was not that Papa was in the war (actually he was not even in the war but perfectly safe in Czechoslovakia, and as I knew from the summer holidays, exceedingly comfortable) but that he did not have a job. Did that mean we were poor? It was difficult to tell, we lived in Grandfather's house, Grandfather earned money, our life had not changed. But obviously to outsiders it all looked different, we obviously were poor as far as they were concerned.

There had been other falls from grace. During the last term, when we had still been in the old building on the other side of the park, a boy and a girl had suddenly appeared. They both wore the uniform of the 'Hitler Jugend' and they talked to us separately. Actually it was more like a cross-examination. A cross-examination in which the original Convent school pupils fared worse. Hating cross-examinations of any kind, and doubly burdened by the mysterious secret of Grandfather's mother and Papa's membership of the Party, I retreated further and further into myself. I just mumbled 'yes' and 'no'. Yes, I was ten years old. No, I had not joined the 'Hitler Jugend'. Why? I did not know. Hadn't my father told me that I was to register immediately after my tenth birthday? My father was in the army. (Here I scored a definite point). Had he not been on leave? No. (This was true in a way though I had of course seen him during the summer holidays.) What about my mother. She was not well. Yes, I had been a pupil at the Covent school. No, I did not know why my parents had sent me to the Convent school instead of a state school. No, nobody at home talked ill about Germany and the Fuehrer. No, never, I was quite sure about it. Yes, I was Aryan. No, nobody ever went to church on Sundays and yes, I would be at the 'Heimabend' in the autumn once the new term in the 'Gymnasium' had started; yes, every Tuesday.

I loathed the 'Heimabend'. Not for ideological reasons. I had no idea what those could be. I simply resented the brainwashing, just as I had resented the brainwashing in the Convent. Only this one was less subtle and there were executive powers attached to it. A complaint from the 'Hitler Jugend' could result in being expelled from the normal 'Hitler Jugend' and sent to something called 'Pflicht Hitler Jugend'; this automatically meant being expelled from the 'Gymnasium'. I had not been particularly enthusiastic about going to the 'Gymnasium' in the first place, mainly because it had been Mama's plan. But now I knew I could not face the humiliation of having to go to an ordinary school. Besides, Mama's wrath, if I received such a punishment, could not even be imagined.

The 'Heimabende' lasted from 4pm - 6pm every Tuesday. We mainly sat around a long table, knitting socks or scarves for soldiers and the Leader, a large blond girl who lived in the same road and was already sixteen (and therefore in the 'Bund der Deutschen Maedchen') would read out patriotic stories to us, or we would sing. This was the first part - the less harrowing one. The second part required that we talked to each other or played games.

I felt a complete outsider. This had so far not bothered me but now I began to fear I might betray my secrets (though I still did not quite understand what they were). The more I tried to think of something appropriate to say, the more difficult it became. In addition, there were all those Sunday meetings, endlessly standing around, listening to speeches and finally holding your hand stretched out and singing the National Anthems and the 'Horst Wessel Lied' (which I never quite understood). As during mass in the Convent, where we had to go without breakfast for the sake of the Communion, I started to feel sick and a few times I fainted. The new order took this with as little sympathy as the previous one, suspecting, actually quite wrongly, that I was play-acting, or (and this was a new one) that I did not have the stamina that befitted a German girl. We also had to wear uniforms. Mama made one for me.

Around this time she began to make clothes for friends and, later, also for acquaintances - for money. This was actually strictly illegal. Only a properly trained and registered dressmaker was allowed to earn money in this way. There was now an additional secret to keep. I had always been good at keeping secrets and one more did not bother me.

Two positive things happened at this time: I met Anna and I discovered Karl May. I am sorry for children who have to grow up without ever reading Karl May. For nearly four years my whole life was coloured by his stories. The ones about the American Wild West, with Old Shatterhand (a part I took for myself) and Winnetou, the Apache chief - and the others in Arabia, where he went with his Arabic companion to Mecca. There were actually many Native American and Arabic words I learned without ever knowing it. (Sometimes I wonder whether my ability to go into long and secret dreams in times of trouble goes back to this time.)

Anna was the first 'other person' in my life. Irma's appearance had been too short, just a flicker of light across the screen, gone before she had established her existence. Hans did not really qualify. But my relationship with Anna opened up the possibilities of including other people in my life. It made me aware of the existence and the possibilities of establishing relationships.

It all began quite simply. I had dragged my feet during the sport class. We had now as many as seven sport classes a week, mostly ball games, and since I had no ball sense whatsoever and did not appreciate the importance of competitive games (after all what did it matter who won a ball game, it was nothing real), I just went along. This time the teacher finally lost her patience with me and made me stand in a corner for the rest of the class. I was sizzling inside myself with fury and humiliation, hearing all the whispers and giggles behind my back. Nobody seemed to have anything better to do than to sneer at me. Eventually the class (and the school day) came to an end and after another telling off, I was allowed to leave. I got into my clothes, put my head between my shoulders and stalked out of the school building.

A girl was walking besides me. She kept on walking besides me even after we were in the street. I shot a quick glance at her and realized that she was the tall girl who, at the time of the entrance examination, everybody had mistaken for being at least fourteen. I expected her to make a remark about my punishment, either sneer (which would have been easily countered by extreme rudeness), or offer her sympathy (which would have been absolutely insufferable). But she did neither. She just informed me that her name was Anna. Was mine Emma? No, I said, furious for being mistaken for a particularly colourless girl but at the same time mollified by her tact. My name was Albertine. 'Ah,' she said. 'Now I know who you are.' And after a pause, 'You live just round the corner from where I live. We can go to school together if your mother stops fetching you.' I winced. Mama had at the beginning still taken me to, and fetched me, from the 'Gymnasium', which had earned me a lot of giggles from the other girls. But fortunately she had suddenly stopped. I said: 'I have told her to stop,' feeling rather grand about the way I had said it. Well, she said, that was good. So we could from now on go together.

I shot another quick glance at her. She was at least a head taller and she carried her school satchel under the arm, whereas I (unspeakable humiliation) was still forced to carry it on my back like a little girl. I resolved, silently, that from then on, I too would carry it the way she did. As soon as I left home, I would take it off and put it back again when I turned into our road. Anna's hair had already been cut to shoulder length; it was no longer in plaits like mine. I thought she looked good. I felt no jealousy about this. Beauty was something I had so far not associated with myself or even considered particularly desirable. But I envied her grown-up manners and those I knew I could copy, if I tried. She seemed, and for a long time this feeling would remain with me, much older than I, more experienced in life.

Neither Anna's mother nor mine viewed our new friendship with much favour. Both considered the other a bad influence on her precious and, so far, strictly controlled daughter. (And in a way, I suppose, they were right.)

Sex, at this stage, held no great fascination for me. I had received my initial information during my first school year, and though I had not believed it, subsequent information here and there made me wonder whether the red-haired girl had not after all spoken something near the truth. Strangely, it was the colourless Emma who proved most informed and most interested in the subject. Her attitude was one of sneering curiosity. I wanted to know, simply for the sake of knowing and I also did not want to belong to that ignorant group who, as everybody said, 'did not yet know.' In the first year at the 'Gymnasium' I had actually had, without knowing it, my first sexual experience. It happened during one of those endless sports sessions. We had learned how to climb ropes; something I could do reasonably well (quite a surprise since most exercises, like jumping over the wooden horse or working on the iron bar, were absolutely beyond me), and had from there progressed to wooden poles. I managed to reach the top of the pole but then I suddenly had a strange feeling between my thighs, pleasant and unpleasant at the same time. It frightened me because I did not quite understand what it was but I had a feeling it was not something one could discuss with the grown-ups. I did not associate it with sex at all and it never occurred to me that it was something one could bring about at will.

Anna's approach was entirely different. She was in many ways (or so I thought) already a woman. Not just because of her figure. She wore a brassiere, something I envied because it brought her nearer to being grown-up. But it was something she bore with shame, typical of the initiates who know there is no way back. It was, however, mainly the way she talked about boys, in whispers, indications and secrecy. Somewhere there was also the son of friends of the family, much older, with whom she had 'an attachment'. I did not learn about the attachment until much later because she treated it as something demanding absolute secrecy, something that did not allow any questions on my part. Even his name I was not to learn until several years later, she kept it like a magical secret and referred to him simply as HE.

She never said (the code to which I had not yet progressed and which I therefore could not understand) whether she had already been kissed. The possibility was something quite beyond my imagination. To be honest, as far as I was concerned, I was not terribly interested but I felt admitting this would have lowered me in her eyes, put me back into the ranks of the ignorant ones she despised. So, just to show my understanding, I did not ask questions but answered her indications with equally knowing half sentences and wise nods, and most of all (when I really did not know what to say) a mysterious, smiling silence.

Only once was my ignorance nearly exposed. I think we were about twelve at the time. She came during the afternoon to our house and whispered she had just seen something absolutely dreadful in a shop window on our way to school. I asked her what it was but she covered her face with her hands and said, it was so dreadful, it could not be mentioned. Had I not seen it myself? I said no, and she just shook her head. Next morning on our way to school she whispered: 'Quick, look!' and turned her head away. I looked, but I could not see anything that would warrant a second glance. 'There,' she said with great urgency, pointing her finger at the figure of a little mermaid with nude top, whose fishtail began just in the middle of her thighs. I still did not see what all the fuss was about. 'You are already spoiled,' Anna shouted and rushed off in disgust. I followed her, still mystified. Part of me felt flattered that she thought me 'already spoiled' (though I had no idea what she meant) the other part felt confused. But Anna refused any further discussion on the subject and, so as not to endanger our friendship, I asked no more questions.

A year later, while reading the *Decameron,* a book Anna had 'borrowed' from her father's library, I came across a sentence in the story of the lost maiden and the hermit in the desert. The sentence read 'and then it came to the resurrection of the flesh.' Together with the homily by the hermit who told the maiden that since god had given him a devil and her a hell it was a pious act to send the devil into hell, it provided a sudden illumination. I felt frightened but I also understood now quite clearly how the act of sex was committed. . But I still was not sure about how the baby (I was of course convinced that one act of sex was enough to produce a baby) would leave the mother's body. The obvious and crude answer was the same way it got in. Here I stopped my fantasies. It all seemed still too much for me.

Hans took Anna's appearance with bad grace. In the beginning I did not notice it,, because I was so used to treat him as something that just happened to be there, without ever considering that he might actually have feelings. Our discussions had always been practical, relating to whatever war we were just fighting and since the advent of Karl May to Red Indian stories. They were never about feelings, so it did not occur to me he might have feelings about the way Anna and I played together, discussed things. Though neither I nor Anna were allowed to invite each other officially, we did sometimes, after one of our long walks back from school, sneak into our gardens, never into the house, since this would have demanded special permission, rarely granted.

Hans began to make fun of Anna. He called her ugly, fat, old and eventually settled on the name 'the long-nosed one'. I just ignored it. I had no wish or inclination to let him influence me in what I wanted to do. However, the situation came to a head one afternoon, when Anna and I were sitting outside our garden shed where the previous year Hans and I had built a bunker. Hans appeared on his balcony and shouting 'long-nosed witch', he threw a sizeable piece of wood, which hit Anna squarely on the head. Anna was horrified. 'Will I have a bump on the head, will it show?' she asked. 'My mother will kill me if it shows, she always says if I am deformed, nobody will want to marry me.' (This was true. Anna's mother had taken her precautions so far as to get Anna involved in some complicated intrigue, only to avoid having her vaccinated against small pox. She felt a scar wherever on Anna's body would seriously impair her ability to find a suitable husband. Years later, when Anna had to accompany her first husband to Cairo, this would have serious consequences.) Mama, who just at the crucial moment appeared in the garden, had witnessed the incident. Hans had in the meantime wisely departed. Anna left for home without any visible harm to her face. In the evening, I heard Mama say to Grandmother: 'He was just jealous', referring to Hans. I was amazed and also furious about the way grown-ups always saw everything in their own perverted way.

Anna and I had long discussions on our way to school. We talked about the books we had read, religion, philosophy, what we wanted to do - she marry, I travel. Many years later somebody told me, he had once walked behind us and listened to our discussions. Anna was than still almost a head taller than I. She carried her satchel under the arm, mine was strapped to my back. I had two thick blonde plaits, broad cheekbones contrasting not too favourably with Anna's slender face. He said we had solved all the problems of the world, practically, metaphysically, and he had, with amusement, watched those two schoolgirls, one talking down, the other up, solemnly putting the world in order.

Whereas Anna introduced me to sophisticated discussion, reflections about the future, and an interest in clothes, I introduced her to extended walks from school, climbing over snow heaps, or hanging around in the Hammer Park near our house. Also to dirty and torn clothes, throwing snowballs or stones at boys if they threw some at us. Once Anna totally startled a raffish looking youth, who had thrown a snowball at us, by drawing herself up to her full height and shouting: 'The Lord in Heaven will punish you.' He stared dumbfounded at us with open mouth but I did not want to leave the punishment entirely to the Lord in Heaven and threw a piece of ice into his face. Next day, his mother rang the bell and complained to Grandfather about my behaviour. But he just smiled and told her to teach her son to fight his own battles and wasn't he ashamed to be afraid of a girl.

And then something really dreadful happened. Nick died. I cried for days. The vet who had come to put him to sleep in the basement of our house was annoyed at the violence of my grief. 'There is a war out there, 'he said. 'People are dying. Don't you understand that? And all you do is cry your eyes out for a dog.' But it was not just a dog. It was Nick, the friend of my childhood, the one who had always protected me, never failed me. I felt a whole area was coming to an end; nothing would ever be the same again. The fairy stories were ending, the mythological area closed. I could no longer be sure that the great big wolf was my exclusive guardian, designed to watch over me, defend me. In future, I felt with sickening fear, I had to defend myself. No Nick between me and the lion of my nightmare. No more rebuttals, it would now all happen to me - directly.

Anna's parents were staunchly opposed to the Nazi regime. Her father had lost his job since he was not 'politically trustworthy'' and he was also not allowed to join the army. Anna's grandfather, who lived with them in an arrangement similar to our own, had a rather Jewish appearance and there was from time to time some gossip, but it was wholly unjustified. They had nothing to hide; we apparently had many things to hide. Nobody knew the story about Grandfather's mother but then, Anna was also not supposed to know of Papa's membership of the

Party. But I was used to my secrets and hiding them presented no burden to me. Somehow, I think I got used to showing only parts of myself to different people. Mama grew more and more bitter. She never missed an opportunity to drop hints about some men being cowards whenever there was an opportunity, having completely forgotten that the heroism she expected of Anna's father was exactly the one that had turned her against Papa.

There is a curious blank that goes right through me and it has always extended itself to music. Now there was a brief interlude where I almost learned to penetrate this no-man's-land. Our music teacher was young and (I thought) beautiful, engaged to an officer who wore a black uniform and who, when in town, would sometimes fetch her from school. She started to teach me the flute, insisting that I could play music if I went about it the right way - and she was going to show me that way. Anna was musical. She had already been taken to the opera, an event that had deeply affected her. Like boys and sex, music was something where she was far ahead of me. Now I began to play the flute in our school orchestra and Frau Doctor even taught me to sing - in her home, where nobody could hear me. Slowly I felt a new confidence growing inside me. And then it was suddenly all over. Frau Doctor was arrested during a music lesson, there were whispers that she had paid a subscription to some anti-Government fund, and that her lover had informed on her. We never saw her again or heard anything about her. I withdrew again into myself, left with a flute I did not know how to use, no more singing lessons. Like Mater in my first year at school, like Irma, as soon as I liked somebody they disappeared from my orbit altogether. I told myself that I must learn to take such things easily. I felt a certain confidence that as I grew older it would become easier to control my emotions, and eventually not to have any emotions at all. At least one had to learn to pretend one did not have any emotions because then (this I knew already quite clearly) people were robbed of the ability to hurt one. Eventually, with enough practice, it might become a fact. (Sadly, it almost did.)

(There is another blank. An inability to feel or show sorrow at real tragedies - or at what most people would consider real tragedies, like death. If people close to me, like Aunt Maria - and later others - suddenly die, I just freeze up, unable to say or even think what I am really feeling, unable to express feelings at all.)

It was Emma who told me about menstruation. I listened with interest and went straight to Anna to show off with my new knowledge. But, to my bewilderment, Anna threw her arms on the table, put her head on it and began to cry. 'Stop it, stop it,' she cried. 'Once you know that you have lost your childhood.' I was totally mystified. Only much later did I understand that she was already menstruating and that her mother had instilled in her a deep shame about the subject. I felt rebuffed. 'Did you know it?' I asked, hurt that she had not told me. But she insisted the subject was too dreadful to be discussed and the least I knew about it the better. I was really hurt. There were obviously areas of knowledge she did not want to share with me. When she thought me 'too young' to be an equal partner, I shrugged my shoulders and accused her of being hysterical herself and, if anybody was a 'baby', it was she, making such a fuss about it. As in the case of the half-naked mermaid, she doubted my good character, which made me feel slightly better.

In any case we did not talk about the subject again and there were a few strained weeks between us. We often had such strained weeks now and they made me feel unhappy. I was jealous of her other friends and the thoughtlessness with which she sometimes treated our friendship (not bothering to wait for me in the morning for days without explanation) hurt me. I affected equal casualness, it is worse to be discovered suffering than actually suffering.

At about this time I was beginning to lose my so far inexhaustible capacity for being alone and it would never again return. Now being alone began to depress me. Hans' company became less and less satisfying; there was simply no relationship. There was another girl with whom Anna had been friendly in her previous school. Her name was Sophie and she became a kind of appendix to our games. Her company never made me quite as happy as Anna's, though she was much more accessible and she shared in some (though not all) our private games. She was also much more 'sensible' than either of us. Her father, who owned one of the biggest pharmaceutical concerns in the town, had already marked out her future. Whereas our future was an adventure with many possibilities, Sophie's was already certain.

Sophie was altogether different. Whereas Anna and I repeatedly got into trouble with our teachers (according to her mother because of my bad influence), Sophie was very much a model pupil, which slightly lowered her in our eyes. She would clean blackboards, carry teachers' belongings and equipment into the classroom, took responsibilities for all collections and, on top of everything, she was an enthusiastic 'Hitler Jugend' member. Her father was determined that she should study pharmacy and heaven help her if she was not capable of doing so. He was also a member of the Party. So was Papa but following Grandfather's instructions I had become skilful in keeping quiet and nobody knew about it. Whereas Anna and I tried to avoid the 'Hitler Jugend Heimabende', Sophie now came to collect us.

Mama and I now spent every summer with Papa in Czechoslovakia. There, for the first time, I was given almost complete freedom. Mama seemed to have her own friends and, except for meal times, I was more or less allowed to roam around. Allowed is perhaps the wrong term, nobody seemed to bother. It was also a time for Mama and Papa to be together without the stifling presence of Grandmother. I suppose they were happy but I just do not know. Mama would never have gone so far as to admit she was happy. I remember how once, she had expressed great surprise, when somebody told her she was happy. 'But one never knows when one is happy,' Mama had said, bewildered. 'If one knows it at all, then it is only after it has gone.'

My enthusiasm for the countryside grew. In Aunt Maria's place I had more or less been restricted to the house, the stables and the immediate surrounding. But here the whole village, all the fields and woods around it were open to me. I started to fall in with a gang of village boys. We played rough games, Red Indians and something they called 'fox in the lair' that involved being beaten with a stone tied into a handkerchief until you got back into your lair, drawn as a circle at a street corner house. I had nothing to do with the girls; I shared none of their interests, neither in dolls nor in clothes. There were some skirmishes where I knew the boys tried to indicate I was female, like grabbing me quickly in a certain way but I just pretended I did not understand.

I still repeatedly read all the (sixty-five!) Karl May books and playing Red Indians was still my foremost interest. If Mama and her friend (whose husband was, I think, in France) went for a picnic, I disappeared in the forest and made, what I fondly believed to be an unseen Red Indian approach. The country around us was beautiful. There were pine forests, unexpected little ponds, large stones between the trees, covered with moss and enormous blueberries, bigger and sweeter and more blue than anything I had ever seen. As in Aunt Maria's place, I felt I loved the country. But now I also felt that I had to love it while I could, because one day I would have to go into the city and do - I did not quite know what. It was a time heavy with destiny and it deeply affected me. But this made it all the more sweet, gave the time in the country a feeling of happiness that I knew would never return. I had to taste it now, every moment, more deeply, more fully as long as it was there. Because one day it would all be gone and all I would have left was the memory and it depended on how I felt now.

I did not much like the son of Mama's friend. He was fat and he wore glasses and there was something funny about him. Sly. One afternoon when I was alone with him, he said he would show me a book. It was full of drawings of couples having sexual intercourse in various poses and the text that accompanied the pictures left little to the imagination. Herman watched me carefully while I looked at it, and I knew if I faltered and betrayed any unease, he would use it to torture me. How I was not quite sure, but it would be something unpleasant. So I just flicked through the pages with a bored expression and said I had seen the book before. I must have convinced him because he looked disappointed and never mentioned it again. There was of course no point in telling this to Mama. Mama would just have cried, rung her hands and blamed me. I was beginning to realize everything was easier to bear without Mama's hysterical lamentations that always put me into the wrong.

It was during this summer, when we were staying in the flat of Mama's friend (Papa had gone to Prague for some reason and Mama did not want to be left alone) that I menstruated for the first time. Emma had explained it to me and other girls had mentioned it too, though Anna had never talked about it. I was neither shocked not ashamed, the only problem was how to tell Mama without letting her know that I already knew about it. This would only have provoked a long interrogation about by whom, where and under what circumstance I had been told. This would then end with the prophecy that I would come to a bad end, and if she had not given birth to me at home instead of a hospital, she was sure I was a changeling because how did she come to have a child like this.

I knew I had to tell her something because if she found the bloodstains in my underwear the same kind of interrogation was inevitable. So I just mumbled something about having a pain in my tummy and after the suitable interval of a few hours, I said I was worried I was ill because there was blood in my trousers. Fortunately Mama was sufficiently embarrassed not to probe. She just said that was all right, it would now happen every month (which I knew) and I had to be careful from now on because now I was a woman. The last statement made me feel uneasy.

I was put to bed with a warm water bottle. Mama and her friend whispered and there was Herman hanging around with a sly expression in his watchful myopic eyes. I had an uneasy feeling he knew what had happened. This more than anything else made me really angry but I remained silent. And then Mama and her friend went out and left me alone in the flat with Herman. And then the implications hit me. If I now was a woman, I could become pregnant. This really frightened me. To become pregnant was one of the offences that could neither be hidden nor denied, it grew inside you without you being able to stop it, everybody could see it. The horror of childbirth and, even more, of having a child around you for the rest of your life, suddenly hit me.

I hope they invent something, I thought in panic. Something like a pill you can take and no matter what you do you won't become pregnant. But it had to happen soon, I was already a woman. I wondered whether one could become pregnant by accident, touching something a man had touched. This really worried me. It removed your free will and your free actions even further away from your control.

I felt more and more uneasy about Herman. Would he try to do something while Mama and her friend were away? (Do what, exactly?) Actually, I felt fairly competent to repel him physically but than one might get pregnant by just being touched by a man. I lay in bed without moving. Herman stayed in the living room. The atmosphere was thick with tension. Once he opened the door and asked, with a sneer on his fat face, whether I wanted something to drink, but I said, no, all I wanted was to sleep. And then Mama and her friend came home. I was amazed at Mama's stupidity. There she had been mumbling threats about having to be careful if I did not want to get into trouble and the next thing she did was leave me alone with that sly creature Herman.

Anna's friendship introduced me to combativeness. So far there had been no need for it. As an only child I had never doubted that I was absolutely unique. There was nobody with whom I could compete. I had never thought about the possibility of having to measure myself against somebody else. Now there was Anna. I wanted her to do well, be appreciated, but at the same time I did not want her to do better. Whenever our exercise books came back, or there was an oral examination, I felt nervous about my own results, but whenever Anna opened her book I felt equally uneasy. I did not want her to fail, to look ridiculous but I was equally afraid she might do better than I.

Anna's mother had, since the day she was born, not only decided she would make a fabulous match with a worthy and wealthy husband, but also that she was a genius, though at the same time she did not want her to be the kind of genius who might frighten away prospective husbands. As a rule, Anna was never allowed to join any sport camps. Whenever we went on a skiing course there was always a doctor's certificate, stating she was not well enough, though she was in fact as strong as a horse. Being 'delicate' was something her mother considered befitted a young lady. She also had to learn everything we learned a year earlier with the help of a private tutor, and there were other private lessons, as for example in French. French was not taught at school but Anna's mother felt it a necessity for a well-educated young lady. Therefore initially, whenever we started a new subject, she was well ahead of us, only to be merciless humiliated by her mother when we all reached the same level and Anna, though still good, was no longer outstanding.

Mama now tried to make clothes for people in earnest. She was actually singularly unfit for it. Not because she lacked talent, she certainly was a very good needlewoman, but she found it totally impossible to discipline herself. Or even organize herself. Since she also never said 'no' whenever anybody brought material, she was soon inundated with orders, giving deadlines she could never (and I don't think intended to) keep. This led to scenes and quarrels and more opportunities for falling ill, have week-long migraines and generally feel sorry for herself. I was more and more becoming the scapegoat. Just as my birth had made it impossible for her to leave her parents' house and establish an independent life with Papa (this became the established version) now she had to 'support herself and her daughter'. Her daughter, who on top of everything was going to an expensive school, though throughout the Nazi years we never had to pay any fees. Slowly I began to cringe inside myself whenever she gave a long lamentation to friends or even to perfect strangers; if the occasion arose. I felt ashamed for her. Showing oneself in a position of weakness did not seem in my eyes the best way of presentation. But I don't think she really felt humiliated by the situation. In her mind she was still the little girl, the undisputed princess. The princess might be in temporary distress but this altered nothing in her inherent nobility.

One of the side effects fell on me. I was now sent to answer the doorbell whenever somebody called on her. I had to stand in our front garden and listen to the long complaints about Mama, with which I secretly agreed. Actually, if she really just needed money, there was a perfectly legitimate way of earning it. Women were asked to work for the war effort against proper payment. But Mama stubbornly refused; using the clause that she had a child under fourteen. Eventually, in 1943, this no longer counted, not even the doctor's certificate that she was suffering from 'a stomach complaint' (unidentified) changed the situation. She was told to work for three days each week in a tailoring shop, specially set up to make clothes for soldiers. On those three days I was supposed to clean the first floor and make the beds. Whenever she came home, she would burst into tears and shout at me about the way I had done my job. Much as I tried my efforts were never acceptable. I just could not see what difference a fold in the middle of the bedspread would make to humanity in general, and to Mama's health, in particular, since she herself was anything but tidy. However, whenever she came home there was at least half an hour of shouts, complaints and prophecies that I would end up a slut, or even worse, and what thanks would she ever get for all the sacrifices she had made for me.

There are not many things I remember from this period. One ended rather shamefully and I still cringe when I think about it. During one of our lunchtime excursions we discovered that the door to the large attic of our school building was not locked. We had never been there, so we quickly slipped through the door and began to explore. It was a large dimly lit room, totally empty. A few ladders were kept in one corner. We put one against a window in the roof and I climbed up and from there on to the top of the roof. I did not know from where the idea came but suddenly it all looked, not only simple, but also rather exciting. Sitting like a rider on the top beam I slid along the roof, gratifyingly followed by the horrified shrieks of the other girls and Anna's 'Don't be stupid, come back.' I ignored them. I carefully went forwards (the 'Gymnasium' was ten floors high) towards the next window. It did not seem at all difficult. I climbed down and let myself carefully drop through the window. Everybody was speechless with admiration, apart from Anna who kept on saying 'I don't want to look, I don't want to look,' covering her eyes with her hands. Proudly, followed by a group of speechless admirers, I went down the stairs and opened the door. And there stood our director and, without a word, he slapped me twice across the face.

I was totally crestfallen. All the admiration gone in a second. How had it happened? Later I found out that some busybody had seen me on the roof of the 'Gymnasium' and had promptly telephoned the school.

The second incident is even more vague. I am not at all sure I really remember it the way it happened. Or what it meant. But one evening, when Mama came back from work, she quickly went to Grandmother. I heard her whisper 'those children' and 'dirty ... begging for food.' And then she said quite clearly 'If we lose this war, they will do it to us.'

Then came the day of the accident. I am still not quite sure about it. When I came home from school (rather late) there was a row going on in our house between Grandmother and Mama. This was nothing new. The only unusual thing was that the row went on upstairs and that it should take so long. Grandmother hardly ever went upstairs; even less seldom did she stay there.

I tried not to listen. By that time such rows (always when Grandfather was away) were becoming more and more difficult to bear. They always started in the same way. Grandmother went upstairs into Mama's room and made a seemingly innocuous suggestion. Mama would flair up, answer with a flood of justifications and accusations, Grandmother would go on, Mama would eventually burst into tears and Grandmother (still sounding perfectly reasonable) would leave the room, sighing. After this Mama got one of her depressions or migraines - what ever. They were both careful not to shout, nobody could ever hear what was said. By the time Grandfather arrived, all would seem well again.

But as this became more and more frequent, it more or less set my teeth on edge. I carefully avoided listening. Apart from fragments of sentences I could not help overhearing (always referring to incidents years back), I never really knew what was happening.

On this particular day, it seemed to have been going on for quite some time. Suddenly Mama's voice became shrill; there was a hushed scream and then the sound of a fall. Mama? Grandmother? I did not know. I simply did not want to know. I slipped out of the house and went across the garden. Hans was under our paradise tree. I climbed over the fence and we began one of our games.

I tried to wipe the whole experience from my mind. But questions returned. What had happened? Who had fallen down? Mama? Grandmother?

Nobody came out of the house, nobody looked for me, called out. I stayed with Hans. Eventually Grandfather arrived. I could hear him quite clearly. I waited. And then Grandfather called my name. When I reached the house he said quietly: 'I am afraid there has been an accident.'

I said: 'Mama?'

'No,' he replied. 'Your Grandmother had fallen down the stairs.'

'And?'

'I am afraid she is dead.'

The rest of the evening and the following days are a blur. Mama was crying copiously. I carefully avoided looking at her, asking any questions. She tried to give explanations, I tried not to listen. Papa came home for the funeral. Just for one day. I never asked any questions and I was given no explanation. Eventually the day receded into the background of my mind. I am not quite sure how everything functioned afterwards. I remember that Mama got permission to stay at home, stop her work for the war effort. To look after Grandfather and myself, I suppose. But I cannot remember what exactly happened. What she did. She had never really cooked before. I suppose the fact that there was hardly any food available must have helped. Grandfather was away the whole day, he left in the morning. Who worked? Who cooked? Who did the house work? There were no more quarrels, Grandfather never quarrelled. There were also no more 'discussions' and I carefully avoided initiating one. We seemed to have lived in a state of animated suspension.

And then, slowly, we slid towards 1945, and with it, towards a year of terrible clearness. I remember everything, day by day, hour by hour, and word for word.

1945 – Endgame

It really began the year before, at Christmas. Until then the war had just been something one saw in newsreels in the cinema before one could watch the proper film. But at Christmas the tide turned.

At first Papa (who was by then in France) was supposed to come home for a week but there had been some trouble and he would not be able to leave. Mama just threw the letter on the table and said: 'All right, there will be no Christmas this year. If your father can't be with us, we have nothing to celebrate.' I thought this was rather hypocritical considering that, whenever Papa was at home, she spent a good deal of time quarrelling with him.

The school closed and I spent the first days of my vacation just hanging around in my room, doing nothing in particular. I read a German translation of *Beau Geste* and wondered whether, once I was older, I could disguise myself as a boy and join the Foreign Legion. Out of sheer boredom I tried to do some homework, but I did not get very far. I was not quite sure whether I should pretend to feel upset (again) about not having a proper Christmas (which I knew Mama expected of me), or whether I should just ignore it. Then on the Eve of Boxing Day a telegram arrived from Papa. It simply said: 'Arriving on Friday evening at eleven, main station.'

Everything changed. Mama took out her handkerchief and cried, Grandfather looked up from the newspaper and said: 'There you are, one should never give up hope so quickly,' - which she took as a rebuff.

It was decided that we would now celebrate Christmas on New Year's Eve. I asked, rather doubtful: 'Can we still get a Christmas tree?' 'Don't be so childish,' Mama said irritably. 'What does it matter whether we have a Christmas tree or not?' But Grandfather put his newspaper away and simply said: 'Well, we'll just have to steal one.'

Stealing a tree was actually quite dangerous by then. Grandfather could go to jail for it, and I might have been thrown out of the 'Gymnasium'. Now that the war had become what was called 'total', everything was 'peoples' property' and stealing from the people was a serious crime. 'Wheels must turn towards victory' had for long been written across the entrance of every railway station. There were posters showing a dark looming figure, the mysterious 'Kohlenklau', who stole coal carelessly wasted. Other posters warned, 'Beware, the enemy is listening' and there were even more sinister ones, which issued stern warnings to people who did not believe in the 'final victory'. But they had been around for ages and Grandfather did not seem unduly worried. So on Thursday afternoon we took out my old sledge and set out for the little forest a few miles outside our town. For some reason I remember that afternoon quite clearly. The snow was falling, slowly and evenly, just as you see it on a picture postcard. The big white snowflakes really did look almost like stars. They settled on our shoulders, in my hair, on Grandfather's hat. They brushed against our faces, and if we touched a branch they descended on us in a soft white cloud.

This was the first time in years that Grandfather had taken me out. We left the main road and followed the footpath leading between snow-covered fields past a few old farmhouses. I clearly remembered them. These were the places where Grandfather and I had walked when I was still a small child. There was the old deserted house, the Devil's House. A woman had lived there a long time ago. She had been so vain that she watched herself in a large mirror whenever she was taking a bath. One day the devil had come through the wall and carried her off. Grandfather had shown me two holes in the wall, one quite small and another, much larger, through which the devil had taken the woman. As a small child I had always been a little afraid whenever we went past this house, especially in the evening, but I had always taken great care not to show my feelings because Grandfather despised cowards. Now of course, I no longer believed in such stories, or in the devil for that matter.

By the time we reached the forest, dusk had fallen. We cut our tree and Grandfather tied the branches with a string and put the bundle on the sledge. We put a blanket over it and I sat down on top of it so that nobody could see what we were carrying. A cold white winter moon had risen above the horizon, covering the fields between us and the town with shining silver. The farmhouses rang with voices and laughter and from the little village on the other side of the forest came the sounds of the church bell.

This was the last day of my childhood.

Next day the weather changed. During the night the snowfall had stopped and given way to an icy wind, which froze the huge snow heaps on both sides of the road into solid blocks of ice and made the footpath slippery and dangerous. This meant that the footpath in front of our house had to be spread with sand before seven o'clock in the morning. Failing to do so was a punishable offence. Since Grandfather worked in St. Aegyd and had to leave quite early, the task was left to me. Anna came for half an hour in the morning and brought a letter from a young soldier, whom she had met the previous spring, when she had been in hospital with diphtheria. She showed off no end about it and I kept quiet because otherwise she would have thought I was jealous. The year before HE had finally kissed her (or so she told me) and since then she had had a number of flirtations: with our Latin teacher, the neighbour's son and boys she had met on holidays. To save face I had started to invent similar adventures, a feat that was often rather difficult. Sophie had been stupid enough to admit she had never been kissed, and the way Anna sometimes patronized her, was quite appalling. I often felt sorry for

Sophie but I could not very well tell her to lie, because I was not sure she would have been able to carry it off, and then Anna would have found out about me too. In any case, I had sometimes doubts whether Anna really believed my stories. She had lately developed a certain way of smiling to herself whenever I narrated a new 'secret'. But so far she had never actually accused me of lying.

As the day wore on, the cold increased, most streets were covered with sheets of ice and not even the sand on the footpath was much help anymore. At around half past ten in the evening Mama and I put on our boots, wrapped thick scarves over our heads and left for the main station. We also took the sledge because Papa was likely to have some luggage and my sledge was the easiest way of transporting it. Because of the complete blackout, and the icy roads, it was rather difficult to walk and we arrived at the station at ten minutes past eleven. The women at the ticket barrier told us that the line had been heavily bombed during the afternoon and so far there was no news of the train. Mama and the woman had a chat about the bombing, the food, the rationing, and the situation in general. Both appeared very confident in the 'final victory' and I wondered whether the woman was lying too, because at home Mama spoke rather differently. Then a train came from Vienna and the woman got busy checking tickets, so Mama and I left.

We did not really know what to do. It was impossible to go home and return later because we had no idea when the train would finally arrive. Mama decided that we should walk along the Promenade, a tree-lined avenue running around the old centre of the city, marking (as we had learned at school) the place of the old fortification walls.

The cold was terrible. Slowly it crept through my coat, my scarves, my hand-knitted mittens, the soles of my shoes, making me feel numb and stiff. From time to time we stopped and kicked our toes against a tree to increase (as Mama put it) the blood circulation. Or we pulled each other on the sledge, running very fast and stamping our feet on the ground. But it really made little difference. My forehead and my nose were aching and when I pinched my legs I could hardly feel it. Once I proposed to go home but Mama got very angry and said: 'We shall wait until Papa arrives. Who knows, it might be the last time we ever see him.' This was how she always spoke in those days.

At last, at two o'clock in the morning, after almost three hours of waiting, the woman told us that there had been some news about the train. She did not know the exact time of its arrival (in any case she would have got herself into serious trouble had she told us - everything was kept very secret in those days), but she advised us to wait from now on inside the station. Twenty minutes later Papa arrived. I had not seen him since our last holiday in Czechoslovakia. He wore a different uniform that made him look older and strangely unfamiliar. We embraced and kissed and Mama at once asked: 'How long can you stay?' Papa answered: 'Only two days, I am afraid.' Mama's mouth became a thin line and being afraid she might start to cry or have one of her hysterics, I quickly began to talk about the tree Grandfather and I had stolen the day before, and how we would now be able to celebrate Christmas on New Year's Eve. We put Papa's luggage on the sledge and all three of us pulled it along to keep ourselves warm. At home I was sent to bed straight away.

Next morning I woke up at nine which, considering we had not reached home before four in the morning, was really quite early. I slipped out of bed and went across the landing to my parents' bedroom. The door was not properly closed and as I raised my hand to knock, I could hear Mama's voice: '... to send you to the Russian front, now of all times. It's as good as an execution.' Papa's voice sounded tired. 'You can't really say that. I know very well it was my fault. I should have followed the instructions literally. They could easily have put me in front of a court martial. All in all it was generous of them to let me go honourably to Danzig to supervise the transport of the refugees ...' 'Generous! Honourable!' Mama's voice was shrill. 'Don't you understand what sort of people they are? Even now? Why doesn't the Kommandant himself go to the Russian front? I'll tell you why. He wants to make quite sure that he falls into the hands of the Americans when the final showdown comes. In a few weeks, the Russians will be in Danzig, you will help the refugees to get out and you will be the one who is left behind to face the music.' Her voice became pleading. 'Why don't you go to your sister's old place and stay in the country until the war is over. It can only be a question of months now.' There was a short silence and then Papa said: 'An officer does not run away just because the war is lost.' For a moment I was rather proud of Papa. Of course, it was impossible for a man to run away and hide like a coward. Mama should have understood, she should have encouraged him, she should have been proud of him. At least that was what women did in the plays and stories we read at school. Had not Caesar's wife sent him to the Capitol? Had not that mother in Sparta said to her son: 'Come with your shield or upon it.' I knew exactly the answer Mama was going to give and to spare Papa from hearing it, I opened the door and cried: 'Good morning! Happy Christmas and Happy New Year altogether!' and laughing stupidly, I jumped on their bed. Papa laughed but Mama said coldly and without a smile: 'Your father has to go to the Russian front.'

I felt terribly embarrassed. First of all the news was no shock to me because I had already heard it. But I could not possibly admit to having listened at the door. Moreover, I did not really know what was expected of me. If I was too upset Mama would have made it more difficult for Papa, and if I was not upset enough, Mama would have called me heartless and ungrateful. So I just said: 'Oh!' and 'I am sorry,' which even to me seemed a rather poor show of emotions.

Somehow the two days passed. Not exactly cheerfully. Mama went around with a tight face but she made (at least in my presence) no more attempts to persuade Papa to go and hide himself in the country. But she was very bitter about Hitler and the whole Government and Papa had to ask her several times to be more careful. On Monday morning, the second of January, Papa left. We accompanied him to the station and watched the train until it finally disappeared in the heavy snowfall.

January

January passed quickly. Papa wrote one letter from Frankfurt where he had gone to join his new unit. He enclosed a note for me, saying I should take good care of Mama; she was very delicate and not at all well. I cringed when I read it.

The school started on the seventh of January and I at once announced that my father had gone to the Russian front. Neither Anna nor Sophie seemed very impressed. Both had their fathers staying at home. Anna's father worked in the Town Hall, distributing ration books. Anna's mother considered this job a great humiliation but it gave them an opportunity to obtain extra food and clothes rations and sell them on the black market. Sophie's father owned the largest pharmaceutical dispensary in town and was therefore 'indispensable'. Mama, as was her habit, and to my embarrassment, often made pointed remarks to their parents about how some people were able to get out of everything.

School was less troublesome than usual. We had lots of air-raids. The bombers usually flew over our town at eleven in the morning on their way to Wiener Neustadt and other industrial areas. At about two in the afternoon they came back.

The school air-raid shelter was actually no shelter at all. It was the cellar of a big house adjoining the school building and whenever a car passed in the street, it shook to its very foundations. The stairs were so steep and slippery that sometimes even a teacher fell down and then it was always very difficult not to laugh. There was a funny machine in one corner that was supposed to pump fresh air into the shelter, but it was so noisy, we were forbidden to use it.

Alongside the walls stood rows of two-tiered camp beds. I once climbed on to a top bed but, as soon as I tried to sit down, both beds collapsed with a most dreadful crash. Fortunately nobody had been sitting on the lower bed. This brought home the fact that I had put on quite a bit of weight over the Christmas vacations. I have no idea how it had happened. It certainly was not the food, we had hardly anything to eat. Anna found the whole situation very funny and from that moment on did not stop making fun of me. She would say, with mock sympathy, that I should not wear big beads (I did not have any beads) and dresses with a flowery pattern were not really suitable for 'fat people'. As if, with the few clothes rations we got, I could have chosen what to wear. She sometimes made fun too of Sophie, who was very tall and very thin; but then it is never as humiliating if people laugh about you for being too thin. You can always pretend you like it. But if you pretend you like being fat, nobody ever believes you.

A few days after school had started, we got the news of Henny's engagement. Of course, Henny (Hans' sister) was quite old, she was almost eighteen. I had known that there was something between her and Robert, partly from Mama's disapproving remarks, partly from what I had seen happening in their garden last summer. (One had an excellent view from my bedroom window and once or twice I had seen them kissing.) Robert, who was the same age as Henny, had been called up to join an anti-aircraft unit in Wiener Neustadt, and Henny's mother had agreed to the engagement. Mama considered it a very stupid idea, I heard her say that this was just 'inviting trouble', but I was not quite sure what she meant by it. Anna tried very hard not to appear impressed by Henny's engagement. Of course, she did not think much of Henny. Henny had only been to a secondary school and her mother owned a milk and patisserie shop round the corner. Still, Anna took great pains to tell us that lots of boys had told her they would like to marry her once she was grown up. Sophie, who could sometimes be terrible catty without knowing it, said such promises weren't quite the same as getting engaged. Anna pursed her lips, looking for a moment very much like her mother, and began to talk about the mysterious HE, telling us in strict confidence that she was actually 'as good as engaged' and that HE was only waiting for her to finish school. Sophie, who never told lies herself, was very impressed but I shrugged my shoulders and said I did not plan to get married. I did not want to have children and I would much rather lead a life like Theodora of Byzantium.

As the weeks passed we spent more and more time listening to the radio. Though nobody would dare admit it openly, we all knew that since the Allies had landed in Normandy, it was only a question of time who would arrive first: the Russians or the Americans. Many people hoped that our army would be able to keep the Russians out of Germany (we were still not allowed to use the word 'Austria'), until the American troops had arrived. But, day after day, we heard news of yet another 'planned retreat' in Poland, in Czechoslovakia or in Hungary. Some years before we had all been given 'war diaries', little booklets with a map on one side and room for us to record the victories of the German army on the other side. Those diaries were collected from time to time and then given back to us. Now, suddenly, without explanation the diaries were collected and we never saw them again. We also heard of the terrible atrocities the Red Army was committing everywhere, and the German people were asked to defend their honour to the last man, the last woman and the last child. In the evenings, we would lock all doors and windows and listen to the BBC. We had a 'people's radio' like everybody else, built so that one could only hear German stations, but occasionally we got London somewhere between Munich and Frankfurt. After the three sharp raps (that always made me think the GESTAPO had already arrived at the door) the program began. A man told us in very good German that the reports about the atrocities committed by the Red Army were just Nazi propaganda. The Austrian people should have no fear, the Russian soldiers came as friends to liberate us from the yoke of Hitler.

Mama would get tearful and nervous. 'What do you think?' she asked Grandfather.

And Grandfather would shrug his shoulders and say: 'The truth probably lies in the middle. The Russian soldiers won't come as friends. A conquering army is nobody's friend. But I don't believe all these atrocity stories either. They just want to put people in a panic to make them fight a little longer.'

In the last week of January our letter to Papa was returned. Across the envelope was written: 'Delivery impossible.' Danzig had been cut off by the Russians. Papa was trapped.

February

In February came the refugees. Nobody was later quite sure when the first ones had passed through our town, but by the end of the second week, the roads were jammed. They came by car or in horse-carts, in one long unbroken, never ending stream, pushing on by day and night, bringing rumours and whispers, and a strange sort of disorder, difficult to describe. People began to wonder whether the stories about atrocities were really just Nazi propaganda. Why should those people have left their homes to tramp over icy winter roads if they had nothing to fear?

The Leaders of the 'Hitler Jugend' arranged a special meeting in our school to organize a refugee service. Our class was ordered to report each Sunday morning to the town's Festival Hall, now a sort of official headquarters. We all went except Anna, who quickly brought a doctor's certificate to say that she was suffering from a heart condition. Anna's mother was rather good with certificates. I think Anna herself was sometimes quite sorry to be left out, but her mother thought sport spoiled a girl's figure and with it her chances of making a good marriage later on.

The Festival Hall looked rather different that day. The big Reception Hall, with the gold framed mirrors along the walls, was covered ankle-deep in straw, and men, women and children were lying on it, side by side, in a long row. Some were still asleep but others sat around in little groups, whispering, mostly in a foreign language. Somehow they did not look the way people usually do. I mean there was something desperate and inhuman about them, as if they no longer minded anything. They were not so much a gathering of individual people as just one solid mass, robbed of their individuality by the same overwhelming disaster.

Sophie and I were ordered to carry large baskets, filled with food, from the basement to the Reception Hall, where the food was distributed by a group of older girls. The baskets were the same Mama used for collecting our dirty laundry, but filled with little packets of sausages, butter and white bread (delicacies we hadn't seen for ages) they looked unfamiliar. We worked very hard because Sophie was extremely conscientious and once she had agreed to do something she would just do it. At one point I suggested putting a few packets into our own pockets but Sophie said: 'That would be stealing', so I dropped the idea.

During one of our trips to the basement a man stopped us. He was not very young - at least thirty-five, I should think - and his face was unshaven and rather dark.

He smiled at me and said: 'Good morning, Miss.'

Nobody had ever called me 'Miss'. I collected all my courage (it would have been dreadful if Sophie had found out that I was just as shy in front of strangers as she was) and said with as much dignity as I could muster: 'Good morning.'

He took his cap off and stepped nearer. 'I am a group leader,' he said. 'We are sixty people, all from around Warsaw. Two days we have been waiting already, Miss, two days. My people are tired and hungry, you understand, Miss? Now,' his voice sank to a whisper, 'if you could just give me a few full baskets right away this would save us a lot of time. We wouldn't have to queue at the counter, see?'

I did not know what to do. Nobody had ever asked me to decide anything; nobody had ever spoken to me so humbly. He probably thought I was one of the 'Hitler Jugend' Leaders. After all, I certainly looked much older than Sophie, who was never taken for more than twelve. I said hesitantly: 'Well, I don't know...'

But Sophie broke in: 'You must go and ask one of the older girls at the counter. We are only carrying the food up to them.'

The man smiled with contempt and without a further word went away. Though I was rather relieved, I said: 'Really, Sophie, sometimes you

92 Spring and No Flowers

behave just like a child', and was quite flattered when she called me 'irresponsible'.

At half past twelve we were allowed to go home for one hour. We went to Sophie's house to eat our bread rolls (horrible to think we could have had proper sausages and real butter if Sophie had not been so squeamish) and then we started to play ball. After ten minutes Anna arrived and we stopped, because Anna looked upon the game as both childish and dangerous. She used to say that if the ball hit one's nose hard enough, the nose might break and then one would be disfigured for life.

Looking at me with exaggerated surprise she asked: 'You have been playing ball?' and I murmured, rather annoyed: 'Sophie likes it.'

We went upstairs to Sophie's room and since there was no heating, we slipped off our shoes and sat on her bed. We always liked to sit like this, with the blankets and pillows wrapped around our legs. It gave us a feeling of intimacy that we enjoyed, because despite our quarrels, we were really quite fond of each other.

Anna said she had just heard some new gossip. Did we know the girl who lived next door to them? She was already in the eighth class, a rather stupid and arrogant thing with fat legs. Well, she was pregnant. Anna had heard her mother discuss it with a friend and the friend had said that the child was from a German officer. Now the school would probably forbid her to take the 'Matura' and she would have to marry the officer, who in private life was just a bus conductor.

'Because,' Anna said, looking at us with an air of superior wisdom, 'my mother says that if such a thing happens, the girl has to marry the man.'

Since neither Sophie's nor my mother ever discussed such topics with us we felt very inferior and quickly changed the subject. Sophie began to relate the incident with the man who had talked to me, addressing me as 'Miss'. With great satisfaction, I watched Anna's face take on an expression of studied disinterest because she always boasted she was the only girl in our class people sometimes called 'Miss'.

The hour passed quickly and in the end Sophie and I had to run back to the Festival Hall. We made an unsuccessful attempt to persuade Anna to come with us, saying that the whole thing was really quite fun, and for once Anna shed some of her superiority and replied almost wistfully: 'I can't. You know how my mother is.'

We spent the afternoon peeling and cutting never ending heaps of potatoes, not really a very exciting job. Once we had an hour's air-raid warning, but since no proper alarm was given, the work continued. When I reached home in the evening, the road in front of our house was packed with cars and some horse-carts. People in heavy boots and fur caps were unloading some camp beds and bundles, and carrying them into various houses. Mama stood in the front garden with Mrs. Werner (Henny's and Hans' mother) and two other neighbours. All were talking in an agitated manner.

I joined them and asked: 'What has happened?'

But Mama gave me a cold look and said: 'Go to your room, I shall tell you later.'

Hurt that not even in times of emergency, Mama would relax the strict rule of my not being allowed to join in the conversation of the grown-ups, I went into the house.

The hall was full of bundles and, through the open door of our sitting room, I could see that two camp beds had been made up by the window. A man and a woman were standing in front of them and when they saw me they smiled. I smiled back at them and said politely: 'Good evening.'

But since I did not want to suffer another rebuff from Mama I went straight to my room.

That was how the Polish refugee, Mrs. Donner, came into our house.

At the beginning of the week a special order had been issued, making it compulsory for everybody to accommodate refugees. In the afternoon, while I had been with Sophie at the Festival Hall, a group had stopped in our road and a man from the 'Ortsgruppe' had come to every house to see who had spare rooms. In spite of Mama's protests he had decided that we did not really need our sitting room and requisitioned it for the refugees.

The group that settled in our street came from Warsaw. They were all related to each other in some way. The sister of Mrs. Donner came to live with her mother-in-law (a woman who only understood Polish) in the house next to the one that belonged to Henny's mother. Henny's mother had not been asked to take refugees because she had three children, and two aunts, who had been bombed out in Vienna, were already living with her. The most interesting person among the new comers was one of Mrs Donner's uncles, who found a place in a big house at the end of our road. He was quite old and his wife, who was very young and very beautiful, was suffering from some sort of dreadful disease and her arms and legs were completely covered with bandages. They had a little girl with them and an ugly looking Ukrainian woman, named Alexandra. I was very intrigued by this set-up, and as time went by, I succeeded worming out fragments of their life stories. The old man was very rich, he had owned big estates in Poland and he had met his young wife in a restaurant, where she had been working as a waitress. She was an orphan and had been brought up by the Ukrainian woman who, after the marriage, had stayed on and become a sort of servant. I never found out the name of the ailment the young wife suffered from, but I once heard, she had fallen ill shortly after the birth of her little daughter and that there was absolutely no hope of recovery. I told the story to Anna and Sophie but they did not seem very interested.

Mr. Donner, whom I had seen that afternoon in our sitting room, left two days later to join the army in Linz. Though at first Mama had not at all been pleased about the idea of having to take in refugees, she soon began to feel sorry for Mrs. Donner (who seemed to have been even better at suffering than Mama) and after some time we all treated her as a member of the family.

It was in the first week of February that the bombing started. There had been an alarm at ten o'clock and Anna and I had left school. (By then it was up to us to decide whether we wanted to go home or into the air-raid shelter at the school - we invariable chose to go home.) We spent half an hour in her garden, playing with her cat, and then, when the radio announced that a group of American bombers was approaching our town, I had left. Not because I was really afraid but Mama did not like me to stay in the street during an air raid warning. At home there was only Mrs. Donner. Grandfather still went to St. Aegyd every morning, he never told us what he did there but it had something to do with defence. I had managed to reach home five minutes before Mama, which was rather fortunate and saved me a lot of trouble and explanation. As soon as Mama arrived, she took off her coat and went into Mrs. Donner's room. By then we could just hear the strange monotone drone of the bombers in the distance. They came in three separate formations, looking like beautiful shining birds. I leaned out of the window of my room and looked up.

Suddenly there was a new sound in the air. I had never heard it before. It was like a prolonged whistle and there was something distinctly evil about it. The sound grew and grew, came closer, wiped out everything even the roar of the bombers. And then the ground shook, the walls trembled and the glass in the half of the window I had not opened burst into pieces.

Blind with panic I ran down the stairs and into the cellar. Mama and Mrs. Donner were already there.

'They are bombing, Mama,' I shouted, 'they are bombing. Do something about it, please do something about it.'

Mama slapped me hard across the face and said: 'Don't be hysterical. Nobody can do anything. We'll have to wait until it is over.'

I cried: 'But they will kill us.'

And Mama said: 'We all have to die one day. You will just have to get used to the idea.'

I looked at her in horror. How could she be so callous, so absolutely without sympathy? And then a new wave of bombing shook the house and, covering my head with both arms, I buried my face between my knees.

It was soon over. The roar of the bombers ebbed away and became a soft murmur in the distance. The silence that followed seemed strangely unreal.

Mama touched my shoulder and said, more softly than usual: 'Come, it's over. Don't sit here in the damp. Let's go upstairs.'

I followed them upstairs and slowly I began to feel deeply ashamed. I had never been such a coward. During our last holiday in Czechoslovakia I had climbed all the trees the boys used to climb, and even a few some of the boys did not climb. I had won several medals for swimming, I had been on a life saving course and once, for the sake of a bet, I had climbed out of a window and on to the top of our school building. An action for which the enraged School Director had very nearly expelled me. I recalled all those ventures and I could honestly say that not one of them had ever made me feel afraid. My only consolation was that it had happened at home. Unthinkable if I had made such an exhibition of myself in the school air raid shelter. Even Anna would probably have behaved with more dignity, I was sure of it. To make up for my failure I decided to visit the area which had been bombed as soon as there was an opportunity.

After the alarm was over Mama and Mrs. Donner joined the women gathering in the street and when they were all busy talking, I quickly slipped away. One of the women had said that the Schiller School was in ruins. If I ran very fast I could be there in ten minutes. I took the short cut through the Park at the end of the road, ran down a long straight avenue and turned the corner. And then I saw it.

The first thing I realized was that there was no protection against bombs. Whatever they had told us at school and over the radio about bomb cellars and air raid shelters had all been lies. The huge school building, which some time ago had been turned into a hospital for wounded soldiers, was a crumbling ruin. The bombs had torn it apart from the fifth floor right through to the basement. Hospital beds were hanging upside down and the roof, with a huge, Red Cross painted on it lay on top of a rubbish heap, as if it were a toy a child had thrown there and forgotten.

I stood quietly and watched the people in uniform and in civilian clothes digging desperately in the ruins. For the first time I understood that we were all trapped. Not just Papa up in Danzig, not only the people in the Ruhr, who had been bombed almost since the beginning of the war, and the soldiers who fought at the front - but we too. The war was no longer a matter of radios and films, of newspaper articles I did not quite understand and conversations between grown-ups to which I was not supposed to listen. It had become real and we were all in it. All of us.

March

March was a strange month. The old order, that for so long had kept everything together, started losing its grip. Here and there, people began to express their dissatisfaction quite openly. For the first time we heard of underground movements and it was generally known that Anna's mother was paying subscriptions to a Monarchist party, and that Mr. Schultz, who lived in our road, was a Communist. There were, of course, other people who talked more than ever about the 'final victory', and there were strange rumours about some miraculous new weapons that would change the course of the war at the last moment. The GESTAPO made several arrests and a number of people were shot, but nobody paid much attention. People no longer knew what they should fear most: the secret police, the bombing, or the Russians. We were, as the saying goes, caught between the devil and the deep blue sea.

Since the refugees lived in our houses, it was no longer possible to believe that the stories about atrocities were all Nazi propaganda. Evening after evening Mrs. Donner would tell Mama about her own experiences in Poland. At first Mama had sent me to my room but then Grandfather had intervened: 'Let her stay,' he said, 'you won't be able to protect her when the Russians come. It is better if she at least knows about it.'

So I sat there, terrified by the stories and at the same time proud, because it was the first time I was treated like a grown-up. It seemed very important not to show fear, to appear calm and composed, as if the whole thing was some sort of secret test.

As time passed the bombing became more regular. Every morning at nine o'clock the 'cuckoo' would call on the radio, which announced the advance of the American bombers. We had like everybody else, a little map of Lower Austria which was divided into small circles, bearing different numbers. As the bombers advanced a voice on the radio would read out the numbers to us. If the bombers came within a certain distance the alarm would sound and we would go down into the basement. I would sit there and listen for the faint, far away humming of the propellers. Sometimes I seemed to hear it long before anybody else did and every day my fear grew. I used to press my fingers into my ears so that I should not hear the growing roar of the bombers and the awful whistling sound that announced the bombing. Together with the fear came a feeling of utter humiliation and of contempt for everything those clever grown-ups called 'authority.'

All my life I had been afraid of Mama. She had always been very strict.

She would tell me what to do and punish me with sharp slaps across the face if I did not obey (or if she found me out which fortunately did not happen too often). But now she sat there and was just as helpless as I was. All over the country the cellars held parents who could not help their children. Trembling with fear and filled with bitterness I thought: 'If they can't protect us, why should they have the right to tell us what to do? Punish us if we do not obey?'

By twelve o'clock it was usually over and we could go upstairs and sit in the sun until half past two. Then the bombers would return from Vienna or from the Marchfeld and pass over us on their way home. Sometimes they dropped a few bombs, and sometimes they just disappeared quietly beyond the horizon.

Those two hours were all we had. Spring had come early that year. The air was warm and the lilac trees had big bulging buds, ready to burst into bloom any moment. I lay on the lawn and saw how beautiful the world was. I had never before noticed the deep blue colour of the sky, the bright reflections of the sun on the white bark of the birch trees. I had taken it all for granted. I had thought it would go on forever. But now everything was different. I was no longer sure of anything. In two hours the bombers would return. I would sit on the floor in our dark, damp basement, my head between my knees, my fingers in my ears, sickeningly, humiliatingly afraid. In two hours I might be dead, or wounded, my legs broken, my face crushed and deformed, or I might slowly suffocate underneath the ruins of our house. Now I was still alive, I was lying in the sun, I was happy. It was a happiness that hurt, but I shall never, never forget it.

At the end of the second week the refugees were still trekking over the roads. But by then their appearance had changed. They no longer came in cars or in big trucks loaded with bundles. Gradually they had come to look poorer and poorer. Their clothes were shabby and torn. Some pushed little two-wheeled carts, some walked on foot, cadging an occasional lift from an army lorry. Their poverty increased my fear. If even those people whom nobody could mistake for capitalists, and who hardly understood one word of German, preferred to run away, what would happen to us?

One morning the refugees had disappeared as suddenly as they had come. A few days later the first German troops began to drive through our town on their way westwards. One night I was woken by a terrible noise that practically shook the walls of our house. Terrified that I had slept through an air-raid warning I rushed into Mama's bedroom. But Mama was still there and together we went into our front garden and looked over the fence. It was a very bright night. The moon stood over the roof of the Church of St. Joseph and, way off, we could see a long procession of German panzers, moving slowly along the main road. They came one after another; their chains thundering over the pavement, their guns looking like broken flagpoles.

And then people began to disappear. First it was a big Party boss whose house was empty in the morning. Then it was a judge who had, as people said 'made many enemies'. Women, whose husbands held influential positions, hurriedly took their children and went westwards with the retreating army. People who had never harmed anyone, and had never even been members of the Party, panicked and left. And with everyone who went the panic grew. Mama would meet a friend in the street and ask her: 'Will you go away?'

And the friend would answer: 'No, no, we won't. We have nothing to fear, we shall definitely stay.'

And in the morning, they too, had gone.

Sophie's father got in touch with us one afternoon and said he was taking the family to Mariazell and would Mama and I like to join them. Mama said coolly that since Papa was probably a prisoner of war in Russia (if he was still alive) she considered it her duty to stay. There had been rumours about a division of Germany and Austria between the victorious Allies. If we left now we might later be in the American Sector and Papa might never be able to join us. Grandfather, of course, was against running away on principle. He hated panic and chaos and believed that whatever life brought had to be suffered with dignity.

I went to say good-bye to Sophie; we packed her rucksack and afterwards played cards. As a matter of fact we quite enjoyed ourselves. We still did not understand how serious it all was.

In the third week of March, our school was taken over by the army and transformed into a military hospital. The Director delivered one last speech and urged us to show ourselves worthy of our great past. He spoke about the Goths, who had fought at the foot of Mount Vesuvius until the last man was dead, and of their women, who had thrown themselves into the crater to escape the shame of being captured by the victorious Romans. For the first time the word 'Werwolf' was mentioned, but I don't think anybody paid much attention to it.

Then came Easter Week. On Thursday, Mama and I went into town. There had been a three-days break in the daily bombing and we all felt a little more cheerful. On the way home we met Mr. Schultz, the Communist. We had never been very friendly with him and his family but on that afternoon he greeted us and stopped, so we stopped too.

Mama asked him straight away: 'How will it be when the Russians come?'

And he answered: 'In the beginning it won't be very good', a remark which, since he was supposed to be a Communist, rather surprised me.

They discussed the general situation. And then he asked me: 'Are you afraid of the bombing?'

The daily ordeal had undermined my pride to such an extent that I did not mind saying: 'Yes.'

He said: 'It will soon be over. There will be a few bad attacks during Easter but that will be the last we'll see of the American bombers. Afterwards there will be air-raids by Russian planes but they won't do so much damage. By the middle of April the Russians should be here.'

Mama and I looked at him as if he were the oracle of Delphi. We did not doubt his words even for a moment. In the way he spoke there was so much authority that it was quite obvious he was already receiving secret information from the other side.

Grandfather still went to St. Aegyd every day. He came home late, usually too tired to talk to us. Mama no longer had to work in the clothing factory but I can't remember what she did all day. Danzig had fallen into the hands of the Russians and we did not know whether Papa was alive or dead. Mama became more and more strange. She did not pay much attention to me but spent most of her time with Mrs. Donner or some of our neighbours. They discussed all sorts of horrible things and once I heard her say; 'If that is true I'd rather kill myself and the girl ...' a remark that frightened me more than anything else I had heard in the previous few months. Sometimes she would lock herself into her bedroom and refuse to see anybody. If I put my ear to the door I could hear her talking to herself or crying. In a way, I had never had so much freedom. For the first time in my life I could really do what I liked, as long as I did not leave the house during an air-raid. But there was really not much one could do in those days - not much one could be interested in. The Russian invasion lay ahead like a dangerous but inevitable illness, and all we could do was wait.

Our house had a large attic that before the war had been filled with old furniture, discarded evening dresses, paintings, books and toys. In former days, I had spent hours in the dusty quietness, trying on some of Mama's old dresses, or reading aloud poems I had composed. But later when the possibility of bombing became more acute, there had come an order to remove everything inflammable from the attic in case of an attack with incendiary bombs. The 'Luftschutzwart', a very unpleasant woman who lived nearby, had come and inspected each house and some people, who had not cleared their attic properly, were heavily fined. Grandmother and Mama worked a whole week and removed everything. All the dresses and the stuffed animals I had loved so much disappeared. The only thing left was a pile of old books and magazines neatly packed away in an old chest that had been pushed into one corner. I had been told that on no account was I to read these books and I could see Mama really meant it. On several occasions I had tried to opened the chest secretly but I had always been found out, either by her or by Grandfather, and the telling off they gave me was really quite ridiculous. Now that Grandfather was away the whole day and Mama did not take much notice of me, I remembered the books. At first I stayed in the attic to read, but when nobody objected, I took a few magazines to my room. Mama saw them the first evening but did not say anything, so I began to read them quite openly. They were full of love stories and scandals about famous film stars, just the material that had so far been taboo for me, and I quite enjoyed myself. I would read them for hours and during that time I forgot the inevitable air-raids, the Russians, the war and everything unpleasant.

One afternoon when I was lying on the floor of my room between piles of old magazines, the music in the radio suddenly stopped. There was a short silence and then came a woman's voice: 'German boys and girls, from the organization 'Werwolf'. Enemies are entering our country from all sides. They are murdering your fathers and brothers, dishonouring your mothers. German youth, it is to you that the dying nation looks. Unite! Stab the enemy in the back! Strike terror into him! Kill, murder the barbarians who are disgracing the sacred soil of your fatherland ... '

And so it went on.

I sat up. I could not help listening. It was the first time the German radio spoke of defeat as an inevitable fact. I suppose, so far I had still hoped the rumours about some miraculous new weapon were true, and the threat of an invasion would somehow pass us by at the last moment. Certainly Mama and a few other women had said that Hitler was very bad, but they had also agreed that at least we knew how bad he was. We did not know what to expect from the Russians, the Americans and all the others. But now even the German radio that, so far, had threatened that the death sentence was the only fit punishment for cowards who did not believe in the 'final victory', said German boys and girls should unite and stab the enemy in the back. Surely that could only mean –

Just then the door opened and Anna appeared.

I said breathlessly: 'Anna, listen!' pointing at the radio.

Anna listened for a few moments, then shrugged her shoulders and said: 'Nonsense!'

Her smile showed that she considered my agitation quite childish. I don't think Anna was in the least interested in politics. Really I don't think she was interested in anything except boys and how to get married as soon as possible. I asked her whether she did not find the word 'Werwolf' somehow fascinating, but she laughed and said: 'No, certainly not!'

Relieved that the spell was broken, I switched off the radio and we sat down on the carpet and discussed Henny and the mysterious HE and Sophie's departure.

April

Easter was a bad as Mr. Schulz had said it would be. We had several air-raids and the centre of the town was badly damaged. One of Mama's friends got trapped in her office and died the same week. Because all the boys between fourteen and eighteen were ordered the join the 'Volkssturm', there would soon not be enough people left to rescue the survivors. The 'Hitler Jugend' Leaders gave orders that all boys and girls between twelve and fourteen should rush immediately to the affected areas after each attack, to help in the rescue work. Special badges were given to us so that the police would not stop us on the way. There were penalties for disobedience.

On Easter Sunday, my turn came for the first time. Running through the streets, I was caught by another alarm and since there was nowhere to go, I tried to hide under a tree. The street was completely deserted and I was terrified. However unpleasant the hours in the damp basement had been, compared to this stark exposure, they seemed like heaven. Even Mama's indifference and frightening lack of fear were preferable to the long empty roads and their loneliness. Fortunately, no bombers appeared. The sirens sounded again and I continued. For a moment I had been tempted to run home, but I pulled myself together. After all, I wanted to become a great writer when I grew up and great writers had to know all about life. How would I ever know all about life if I ran away each time I was afraid?

I met a policeman on the way and he directed me to the street where our school stood. He said a big block of flats had been hit there. When I arrived I saw a large mound like a gigantic rubbish heap and people running up and down between the ruins like disturbed ants. A 'Luftschutzwart' got hold of me and I was given some work to do. There were a few more boys and girl of my own age but I did not know them. We helped to dig; we carried large beams and stones. When, in a corner of what had previously been a basement, rescuers found a group of people, who were badly shaken but otherwise unharmed, we were told to help, wash the dirt off their faces and find them something to drink. Not all the people found under the ruins had escaped unharmed. Some had to be carried away on stretchers. Some were moaning and throwing themselves about. Some lay very still, strangely bent like broken dolls. Over those people we put a blanket or a coat or whatever was at hand. A woman was still clutching a baby in her arms. The baby was dead; instead of a head it had just a small, ugly, red ball. The woman was screaming. Somebody, who looked like a doctor, tried to take the baby away form her. She would not let him. She just screamed and screamed. She spat at the doctor and tried to bite and scratch him and said Hitler was a murderer and she wanted her baby back. An old man came up to me and told me to go and help somewhere else.

After Easter the sirens remained silent. The Russians had crossed the Austrian border and their planes came now without warning at any time of the day or night. They were small planes and they flew quite low. Their bombs made a lot of noise but they could not really split a house to its very foundations. That thought somehow comforted me.

On Tuesday, Mama and Mrs. Donner went to town to buy food. The day before shopkeepers had been ordered to distribute everything they had in store so that nothing would fall into the hands of the enemy. Soon after Mama had left some Russian planes came. They dropped a few bombs, strafed the street and then disappeared again. I felt a little uneasy. Some of the explosions had sounded quite near.

Hans called out to me from the garden fence and said some people had been killed in Hammer Park. Hammer Park was about ten minutes walk from our house. Together we climbed over the fence and ran down the road. When we reached the Park we could just see an ambulance, driving away at great speed. But that was not all. Something was lying on the pavement. It was covered with a blanket. We edged nearer and Hans whispered: 'Look, it's a woman. You can see her shoes.'

I don't think I had really expected to find anything like this when we had decided to come. The Thing lay there like a huge parcel, a pair of brown worn out walking shoes with wooden soles peeping out at one end. Nobody was near. The roads were all deserted.

Hans said: 'I bet you haven't the courage to lift up the blanket and look at her face.'

Of course, I did not really have the courage. But Hans was one year younger and I had therefore always considered him vastly inferior. Slowly I went closer and lifted the blanket. The Thing was a woman I knew only too well, in fact I had always called her 'auntie'. It is one thing to pluck up enough courage to look at a corpse but quite another to discover that the corpse is someone you have known all your life, who used to bring you sweets at Christmas and books on each birthday.

I was still holding the blanket with two fingers when the plane appeared. It flew quite high but it rapidly moved down in large circles, leaving no doubt about its intentions. I dropped the blanket and without a word we both started to run. All the houses seemed locked and quiet, and in our panic we did not even try to find shelter. We just ran and ran. The plane came down like a huge dragon and when it swooped above us along the road it fired a few staccato shots into the houses to our right. I stumbled over a stone, and for a moment I thought I had been hit, but then I realized I was still running. By the time the plane turned back we had reached home. We jumped over the garden fence almost at the same moment and rushed straight down into the basement. We were frightened beyond words.

By the time Mama returned and Hans had gone home, I had sufficiently recovered to give the story a more heroic appearance. But Mama had little understanding for heroism. She slapped my face and said in future I was not to leave the house on my own. She had enough worries without my adding to them.

For two days nothing happened. The German radio gave no further information about the fighting and, except for occasional demands to form the organization 'Werwolf', remained quiet for most of the time. We burned flags and books, and even some of my schoolbooks, because Mrs. Donner had told us that if the Russians found anything bearing a swastika they would certainly shoot us. (We were very careful to do it secretly. Had we been reported to the police we could have been in serious trouble.)

The next thing I remember was that weird Thursday afternoon. Mama, Mrs. Donner, Henny's mother and I were sitting on our verandah. It was a real spring day, warm and sunny. The laburnum trees at the end of the garden formed a golden screen that hid the houses behind. We were not unhappy that afternoon. Mrs. Werner was knitting something for the baby and Mama seemed more cheerful than usual.

Suddenly Henny came running up the garden, crying: 'Stop knitting, mother. The Russians are coming!'

For a moment we just stared at her. Then Mrs. Werner, who did not really love Henny because she was only fond of sons, asked sharply: 'Why are you not in the shop? What's all this nonsense about the Russians?'

Henny said: 'A woman came into the shop. She asked me 'Do you know the Russians are supposed to come tonight?' When I asked her how she knew, she said, 'My little girl went to town to meet a friend. A policeman stopped them and told them to go home and tell their mothers to pack and leave the town immediately because the Russians were already on the other side of the Traisen.'

The River Traisen was exactly half an hour's walk from our house.

The grown-ups looked at each other in silence.

Then Mama said: 'Oh, God!'

Mrs. Werner fell back into her chair and started to cry: 'Please don't leave me,' she wailed, 'I have three children and two old women to look after. Please don't go away and leave me.'

Mrs. Donner dropped her own knitting and said we should all go away. Immediately!

'It will be dreadful,' she said, 'I know, I have lived through it twice.

They will loot everything, they will shoot the men, no woman will be safe.'

I had heard all this before but now it sounded very differently.

Henny broke in: 'We can't go like this. We need a permit from the 'Ortsgruppe' to leave town.'

I said quickly: 'Come Henny, let's go and get one.'

The confusion of the grown-ups upset me more than anything else. I simply could not face it. It was just as if we were on a ship, and the captain had come, and said, the ship was sinking and he did not know what to do about it.

The 'Ortsgruppen' office was very near the place where Hans and I had found auntie a few days ago. A long queue of women and children were already waiting outside. Like us they had come to get a permit to leave town. Henny, who knew people because of the shop, began to talk to some of them. Nobody had heard that the Russians were already on the other side of the River Traisen but they all expected the invasion that night. Some said Vienna had capitulated, and others said the German troops were just leaving Neulengbach.

We seemed to wait for ages in that queue. At last my turn came. I did not know the man who sat behind the desk. For a moment I felt sorry I had come but then I took my courage in both hands and said as calmly as I could: 'I want a permit to leave town.'

The man smiled at me in a rather unpleasant way and asked: 'For yourself only?'

I replied: 'No, for my mother and my grandfather too.'

'Oh,' he said, leaning back in his chair and surveying me, 'and why, I would like to know, doesn't your mother come herself? Or your grandfather?'

I thought desperately, what will happen if he does not give me the permit? Everybody will run away, leaving only us to face the invasion. So I lied quickly: 'My mother has to look after our little baby, she can't come. And my grandfather is ill.'

The man looked hard at me and then, without a further word, took a piece of paper and wrote our names and something else on it.

I snatched it up and left.

Henny was waiting outside and we hurried home together. Proudly I handed the permit to Mama. I felt, I had saved us all.

We spent the afternoon packing. Mrs. Werner produced a little handcart and it was decided that the eldest of the two aunts, who could hardly walk, should ride on it and hold on to the baby and whatever food we had, as well as our most essential clothing. Now the only thing left was to wait for Grandfather. As soon as he returned, we would set out for the Dunkelsteinerwald, a forest several miles to the west of the town.

The rest of the afternoon passed slowly. Hans was sent out for further news but nobody seemed to know anything definite. The three grown-ups began to discuss the situation. Had the Russians already captured St. Aegyd? Had the railway line been bombed? Who would come first: Grandfather or the Russians? Mrs. Donner wanted to leave at once but Mama refused to go.

At nine o'clock Grandfather came back. He took one look at our handcart and asked whether we had all gone mad. Mama explained that the Russians were expected that night and that we had decided to seek shelter in the Dunkelsteinerwald until the invasion was over.

Grandfather said: 'Until the invasion is over? What do you stupid women think a war is like? A house and a town around you offer you at least some protection. Out there in the forest they can beat you to death with an old branch.'

I had never seen him so angry. He was outright rude with Mrs. Werner and Mrs. Donner but they did not seem to mind. He said the Russians would certainly not come before the weekend. The best everybody could do was to go to bed and get some rest. He did not tell us how he knew the Russians would not come tonight and nobody asked. Mrs. Werner even said: 'Thank you,' and took her handcart, her children and her two aunts and went back to her house.

We spent the evening in a strangely relieved atmosphere. After Mama had gone upstairs I stayed behind.

Grandfather took a book and I said hesitantly: 'Grandfather -'

He looked up and asked: 'Yes?'

'Look here, Grandfather, you know I am always reading books about the Foreign Legion and travels and wars and American Indians. I just wonder, wouldn't it be a good idea to dig a tunnel to Mrs. Werner's house? If somebody came to our house we could crawl through it and Mrs. Werner and Henny could come to us if necessary. We could also make a good hiding place behind the staircase in the cellar....'

I stopped, feeling uneasy. I was certain Grandfather would tell me to keep quiet, books were books and life was something else, something I did not yet understand. But Grandfather said nothing of the sort. He looked amused and then he smiled and said quite seriously: 'Not a bad idea, I would never have thought of it myself. We shall talk more about it tomorrow.'

The world of books and the world in which we lived had suddenly become one.

For two days we worked very hard. Grandfather did not go to St.Aegyd anymore. The Russian troops were coming nearer and nearer and the railway line to the south could be cut off at any moment. In such an event Grandfather would have been unable to return to us in the evening. I think, he was very worried that Mama and Mrs.Werner might lose their heads again and make another attempt to go into the Dunkelsteinerwald. There was some talk about Grandfather being accused of treason (or desertion, I was not sure which one it was) if he did not go to St.Aegyd, but I heard him say that he would just have to take that risk.

I remembered how, as a small child, I had been very fond of Grandfather. I used to say I would marry him once I had grown up. I liked him because he always had time for me. Later, once I had started to go to school I spent less and less time with him, and when the war started and Grandfather had to work all day in the office, drawing houses and bridges and supervise much of the building work himself, I seldom saw him. Now, for the first time in years, my childhood seemed to have returned. Grandfather and I spent the whole time together. We worked and planned and he discussed everything with me, as if I knew just as much about it all as he did. Hans came over to help and we built a neat little tunnel to Mrs. Werner's house. Grandfather masked it with a heap of wood and old rubbish on both sides an,d if one lay down and used one's knees and elbows, it was just large enough to crawl through. The room in the cellar, that had so far been used for washing the laundry, was transformed into some kind of living room. Grandfather put wooden planks over the stone floor and we took mattresses from the bedrooms and blankets. (Mrs. Donner had told us the best way to discourage the Russians from staying in a house was to make everything look as untidy and dirty as possible.) Grandfather even arranged a little stove in one corner because we did not know how long the fighting would last. Then we put heaps of sand outside each of the cellar windows to protect us from stray bullets.

The staircase to the cellar was in two flights. In the space under the lower flight we put a trunk filled with our most necessary belongings and Mama's best jewellery. Grandfather even sealed off the space with some old wooden planks, leaving two of them loose so that they could be moved without too much difficulty. Despite the trunk there was still enough room for at least two people to hide. In the space under the upper flight of stairs, Grandfather cunningly put suitcases and clothes and some cheap jewellery and, in front of it, an old wardrobe. If the Russians became suspicious and removed the wardrobe they would think this was all we had hidden and look no further. The idea seemed brilliant. It reminded me of a history lesson where we had learned about the ancient Egyptians, and how they had hidden the coffins and treasures of their dead kings.

On Sunday afternoon Anna and her mother came to visit. I showed Anna the basement and she seemed quite impressed. She said it looked reasonably comfortable and she wished her father had built something similar in their house. But Anna's mother pursed her lips and said this was not necessary in their case. For them the Russians came as liberators. They had suffered enough under Hitler and now, at last, everything would return to normal. She did not look very pleased when she said it and I think in her heart she accused us of being Nazis because we did not have full faith in what the English radio said.

On Monday morning we heard it for the first time. In the beginning it was just a faint thunder, far away, like a distant call in the mountains. But slowly it grew louder and as time passed it lost its anonymous character and disintegrated into the distinct sound of shellfire. Then it died down for several hours and became hardly more than a soft murmur. This happened again and again. Whenever there was such a lull, new rumours swept through our little town: the Russians would not invade us after all, the Red Army was advancing in two separate columns, one along the Danube and the other along the southern border of Lower Austria. Then an unnatural gaiety would take hold of everybody. We would laugh and joke as if the war was already over. In our hearts, of course, we knew that this was impossible and when, after some time, the thunder of the artillery returned, we were hardly surprised.

Many things happened in that week. The German troops were now rapidly retreating. It was said that a few soldiers, who had tried to run away to their families, had been shot by their officers, and their bodies could be seen lying in the streets with a note pinned to their backs, saying this was the way Germany treated traitors. We did not believe it at first but then Henny, who had been in town, came home and said she had seen one herself.

On Tuesday, Henny's fiancé came for a short visit. Wiener-Neustadt had been captured by the Russians as well as the anti-aircraft unit to which he had been posted, and had been ordered to go westwards and join the fighting troops. He told her the army would make one last stand in the Dunkelsteinerwald and would probably fight until the last man was dead. He tried to persuade Henny to go with him, repeating all the horror stories we had already heard from Mrs. Donner. He said it would be better if they died together instead of worrying about each other. But Mrs. Werner started to cry and said Henny could not possibly go away and leave her with the children and the two old aunts. Henny was very upset and finally Grandfather was asked to decide the matter. Grandfather had a long talk with Robert and in the end convinced him that Henny was safer with us than in the middle of fighting troops.

Henny accepted Grandfather's decision with surprising calm. She merely said she would like to see Robert off at the police station where he had to join his new unit. She asked me to accompany them. I don't think Mama liked the idea but she did not stop me. We walked slowly down St. Joseph's Street, for years my daily route to school. Henny and Robert talked quietly to each other, making plans for how they would meet after the war was over, and Robert promised her that he would not fight until he was dead, but try to escape at the last moment. I felt that in some special way they were far away from me, so I did not disturb them. I just wondered how it must feel to be eighteen and in love, to be engaged and to belong to someone and to see that someone go away to war. I glanced at Henny and, for a fleeting moment, she seemed to look as old as Mama had looked on that cold January morning when we had fetched Papa from the station. A group of soldiers were already waiting and Robert had to say good-bye quickly. We waved until the lorries disappeared and eventually two policemen asked us to leave and go home.

The continuous thunder of shellfire, drawing slowly nearer and nearer, increased the panic in our town. Many people, who had been determined to stay, changed their mind overnight and taking nothing but their children, tried to escape to the west. But the roads were blocked by the retreating army, by refugees and by troop reinforcements driving in the opposite direction. Many people died in the confusion and some lost their children, never to find them again. A number of people committed suicide, just as the German radio had advised us to do. Among them were friends of Grandfather, a nice old couple who had always lived a very quiet live. Their death was a great shock to all of us. Others were shot by the GESTAPO for expressing anti-German sentiments or showing signs of defeatism. We later found out that Mama's name had already been on a black list.

Wednesday brought a very bad air-raid. We spent two hours in the basement. When at last the Russian planes had left and everything had become quiet again, we went out on the balcony in front of my room for a little fresh air. Far away, south of the town, half-hidden behind long rows of houses, a huge fire was burning. It illuminated the horizon with sudden flashes of red and yellow and, long after Grandfather and Mama had gone to bed, I was still looking at it. I remembered a poem we had read at school some time ago. It described how the hordes of Genghis Khan had poured into Europe, the long trail of atrocities that marked their way, and the terror of the people who, night after night, awoke to see 'far away flames like a burning village' - a sentence that for some reason had deeply impressed me.

There in the south something was burning. I did not know what it was. A town? A village? Whether it was deserted or whether there were people in it, men, women, children. Whether they were hiding in safety or fighting for their lives or slowly burning to death. But now I understood the poem. I understood the fear and the horror that had escaped me when I had first read it. For centuries the cry: 'The Huns are coming!' had struck terror into the hearts of people and now I felt I knew what it meant.

On Thursday morning a few army lorries stopped in our road. I think something had gone wrong with one of the motor engines and the men were trying to repair it. Mama and I stood in our front garden and one of the soldiers came up to the fence and asked for a glass of water. I don't think he was very old; hardly twenty, I suppose. He had fair hair and pimples all over his face, just like Hans, and Anna always said that if a man has pimples on his face he is not really grown up.

Mama sent me for the water and when I returned with it I found her talking to him. In those days Mama was definitely not very pleasant. She was upset and bitter and she spoke out and said exactly what she thought. I could hear her say: '... you will all go into the Dunkelsteinerwald and leave us to the Russians.'

The young soldier looked very awkward and said: 'It's better for you people if we don't fight for the town. It will be over more quickly.'

Mama looked at him with contempt and said: 'Is that all the great German army has to offer us? Run away so that it is all over more quickly?'

For a moment nobody said anything and then, suddenly, the young soldier bent his head over the fence and began to cry. 'It's all so bloody rotten' he sobbed, 'I know just how you must all feel. Those bastards, they promised us we would win. I am from Berlin and they say up there every Russian carries a pamphlet. It's written by some filthy pig ... it says they should rape every German woman they can lay hands on ... my mother, my two sisters are up there ...'

He put his face into his hands and his whole body shook.

Mama went closer to the fence and said quickly: 'Come in, don't cry in the street.'

Because if an officer had seen him he would probably have been shot for defeatism.

We took him into the kitchen and made him a cup of tea and for the first time in weeks Mama became almost human. She even gave him some of the cigarettes we had been saving to buy food on the black market, and I told him about the bunker Grandfather and I had built, because I thought this might distract him. He drank his tea and when he laughed he did not look ugly at all but rather nice and cheerful. I made up my mind to tell Anna that he had tried to kiss me, though I would, of course, not mention the pimples. By the time another soldier came to fetch him, we were in quite a happy mood. Mama gave the other soldiers some cigarettes too and we both waved at them as their lorries drove off.

I can hardly remember the days that followed. Somehow they are a blank empty space and if I close my eyes I see nothing but a large board with just one word written on it: WAITING. Waiting as one waits for a nasty examination, an afternoon with the dentist, or, perhaps later, when one is grown-up, for that terrible day when the baby comes.

Something I do remember is my birthday. On the ninth of April I became fourteen. I had always looked forwards to this birthday because fourteen is rather special. Even the meanest grown-ups usually agree that at fourteen one is not really a child anymore. Of course, I knew I would have to go to school until I was eighteen, and as long as one goes to school one isn't exactly an adult either. But still, fourteen was something special. Long ago Mama had promised I could have my hair cut on this birthday. Anna had worn her hair in a beautiful shingle since she was ten, and Sophie was one of those fortunate girls who had never let it grow. Quite a number of girls in my class wore their hair short and it was terrible humiliating to belong to those who had to go around with two long plaits - hated attributes of childhood.

With the Russians and the war and everything else I did, of course, not expect a real birthday celebration, least of all a cake, because there was nothing one could use to make a cake. But I did hope Mama would remember her promise about the hair. The morning and half the afternoon passed and nobody said a word to me. Slowly I began to realize that they had simply forgotten about my birthday. Not that I blamed them, but still, it hurt. I went up to the attic and sat on the old chest. I usually went there on my birthdays and on each New Year's Eve and thought about the past year and made plans and resolution about the next one. But on that particular day, when I sat there with the thunder of the shells and the occasional roar of a stray Russian plane in my ears, I just did not know what to think. Life had become so different in the last few months so that everything that had happened before seemed far, far away. It might as well have happened in another country, or to somebody else. I thought about the year in front of us, about all those frightening stories I had heard and suddenly I felt, not only afraid, but utterly and unbearably miserable. I put my head on my knees and carefully avoiding any noise, I began to cry.

I had not been sitting there very long when I heard voices on the landing. Quickly I tried my eyes and went downstairs. Mama did not like me to sit alone in the attic.

Anna had come. She gave me a big smile and said: 'Happy birthday!'

With great satisfaction I saw a guilty expression creep over Mama's face when she heard it.

I said, with purposeful exaggeration: 'How sweet of you. Please come in,' and ushered her into my room.

Anna took a little box from her coat pocket and gave it to me. I opened it. It contained her silver ear clips. A year ago Anna had received them as a Christmas present, they were real silver ear clips, like those the grown ups wore. The design was a three-pointed leaf with a small half moon. I had admired them passionately, and I knew that Anna was very proud of them.

I said: 'Oh no, Anna, they are your favourite ones. You can't give them away.'

Anna looked very pleased with the effect of her present and said: 'Of course I can. You like them too.'

We looked at each other and embraced and laughed and cried a little and I said over her shoulder: 'They have forgotten my birthday.'

And Anna sighed and said: 'I thought so.'

Somehow Mama managed to bake something resembling a cake. It was made of maize flour and she decorated it with a little jam we had saved from last year. There was even a candle in the middle, in place of the traditional fourteen candles, that would have marked my age. Everybody was happy and I told myself it would be mean to feel disappointed because Mama still had not remembered her promise about my hair.

The next week passed quickly. On Saturday morning the Germans blew up the water works and the electricity plants and, in the darkness, our little bunker lost its comfortable look. It became a dark hole without any proper ventilation, resembling more a trap than a place of safety. Grandfather and Henny went with buckets to the nearby millstream for water. I wanted to go with them but Mama stopped me. Now that the Russian shells were already hitting our town, the streets were less safe than ever. Strangely enough I was not as afraid of the shooting as I had been of the American bombers. It seemed more human, less mechanical; something one could somehow cope with.

A friend of Mama's, who had a house very near the railway bridge over the River Traisen, came with a suitcase and asked whether she could stay with us for a day. She had been told that all the bridges would be blown up at noon, and since her house was likely to be damaged, a soldier had advised her to leave.

Noon came and with it a number of explosions which rocked the house. Barricades and barbed wire fences were thrown across the roads to stop the advance of tanks and lorries. The last German troops left. By evening not a single soldier was to be seen in the streets. The town lay bare and empty, defencelessly awaiting its conquerors.

The day ended, the night came. We were now five people in the dark basement room. Mama and I lay together on one of the small camp beds. The shellfire increased. Mama put a pillow over our faces to protect us should the ceiling come down. It was very hot but nobody got up and nobody spoke. We just kept on waiting, pretending to be asleep. The noise of the artillery was now a constant roar. There were also sounds like that of breaking walls or tumbling brickwork and I was convinced that half our street was already in ruins. Grandfather said I should not worry about the explosions: as long as we heard them we were quite safe. It simply meant that some shells had passed over our house. I felt there was some flaw in this argument. What about that particular shell already aimed at our house, the one I would not hear? In the morning, when the noise became unbearable Mama's friend began to pray and that made things even worse. It was as if we were already dead, imprisoned in the darkness of our tomb, and a distant impersonal voice was reading our funeral service. I don't know whether Grandfather shared my feelings but after some time he asked her to stop.

At six o'clock the bombardment died down and we all dozed off. The interval was short lived. After what seemed hardly more than a few minutes, a shrill howling sound awoke us. I recognized it immediately. I had heard it in the newsreels; it was the noise of German 'Stukkas' attacking. I sat up and thought: 'If our own air force is bombing us, we are now enemy country, the Russians must have taken the town.' Grandfather lit a match and to our great surprise we discovered that it was already nine o'clock. In the hot dark dampness of our little enclosure we had completely missed the beginning of the day.

After the 'Stukkas' there was silence. Only a lonely machine gun rang out from time to time. Compared to the thunder of the shelling it sounded harmless and nobody took much notice of it.

Henny and Hans came through the tunnel and Grandfather said we should go upstairs. Before anybody could stop me, I went to the large window halfway up the staircase and looked out. To my immense surprise I realized that nothing had changed. The road was still there, all the houses I had known since childhood, everything! In the pale light of the sunless morning, the street looked quiet and undisturbed as if the horror of the night had just been the product of my imagination.

Slowly I went downstairs. Grandfather, who believed the best medicine against panic was to stick to one's daily routine, insisted we should make some tea. Mrs. Werner came with the baby and the grown-ups began to discuss the situation. Had the Russians come? Was the fighting still going on in some of the streets? Had the Germans driven the Russians back? What should we do? Should we stay in the house or go out and find out what was happening? Slowly, but certainly, the silence, so welcome at first, began to make us feel uneasy.

Henny, Hans and I went up to the landing and sat on a step of the staircase, from where we had a good view over the road, without being visible from outside. We were discussing the night, the attack of the

'Stukkas' and what would happen next, when suddenly, there was a noise in the road and turning to look we saw a long line of horse carts moving past our house. They were full of soldiers and Henny said: 'They are Hungarians.'

But Hans shouted: 'No, no they are Russians! They have red stars on their caps!' and still shouting he ran down.

To our immense surprise Grandfather slapped his face and told him to keep quiet, the soldiers were of course Hungarians. Slowly Hans came back, holding his cheek, looking very affronted.

'Your grandfather,' he said in a hurt but carefully hushed voice, 'is wrong. I have seen it, they have red stars on their caps.'

Henny and I said nothing, but I think we were both surprised at Grandfather's action. Neither of us had ever seen him lose his temper in such an unreasonable way.

Downstairs the grown ups were still talking. The baby had started to cry and we could hear his shrill wailing over the murmur of their voices. We did not talk to each other; we just sat there and looked out of the window. And then I saw Mr. Schultz. He was walking down the road with another man in civilian clothes. There was something different about the way he walked, all of a sudden he looked important, and, how shall I say it, official.

I opened the window and cried: 'Mr. Schultz.'

He turned and seeing me shouted: 'Get away from the window. The Russians are in town, hide yourself!'

I obeyed quickly. It had been all right to build the tunnel and help Grandfather with the bunker. It hadn't really been very different from putting up a tent on a camping holiday; pretending one was in the wilds of Arizona with a hostile Indian tribe in the next forest. But now the touch of unreality had gone. The adventures had become everyday life and only the grown-ups seemed fit to deal with them.

After Mrs. Werner and the children had departed I sat alone in the basement. I could not find a candle and the darkness did not improve the situation. After some time Henny returned through the tunnel and we decided to sit on the steps. It was not quite so dark there. Grandfather had merely nailed wooden planks in front of the windows and there were narrow gaps between them. Long pale arrows of light squeezed themselves through those cracks and fell down on us, transforming the darkness into a misty twilight.

For some time we did not speak and then I said: 'There is something I don't understand, Henny. People say Mr. Schultz is a Communist but he seems just as afraid as everybody else. Now if he is a Communist why is he afraid of the Russians?'

Henny said: 'He isn't a Communist, silly. He is a Socialist.'

I asked her whether that wasn't the same thing but she said, no, there was some difference. Robert had explained it to her. I asked her what that difference was and she answered: 'It has something to do with property. The Socialists leave you something, but the Communists take the lot.'

We again relapsed into silence. Really, the way the grown-ups arranged the world, the things they called 'ideologies' and were so proud of, there wasn't much to choose between any of them.

Since we had nothing else to do, we listened to the noises outside. From time to time Mama's voice rang out, doors were opened and closed. Once I thought I could hear Grandfather talking in the garden, but I was not sure. Then came heavy steps, somebody was shouting, the door above us opened and Mama hissed: 'Hide yourselves. They are in the house.'

For a moment we both panicked. We managed to get to the hiding place but in the darkness we could not find the loose planks. I ran my fingers over the rough wood but it yielded nowhere. Henny kept on saying: 'Hurry, hurry!' but that made me only clumsier. Through our confusion we could hear Mama's voice already outside the basement door. We ran for the tunnel. Henny went first and pulled me through by my arms the moment steps sounded down the stairs. I wanted to run upstairs but Henny caught my arms and whispered it was better to stay, we did not know what was happening in their house. We crouched on the floor in front of the open mouthed tunnel and listened. Somebody was carrying a light, we could see the yellow glimmer coming and disappearing at the other end of the tunnel, and from time to time I heard Mama's voice. It sounded strangely muffled and I could not understand a word. Once it seemed as if the person with the light was shining it straight into the tunnel and I almost screamed but Henny put her hand over my mouth and stopped me.

I don't know how long we stayed there. When at last Mama called us it was already late in the afternoon. We went upstairs and I saw Grandfather in an armchair and Mrs. Donner pressing a wet towel against his forehead. Mama told us that after we had gone, Grandfather, who spoke some Russian, had insisted on going out despite her protests, to see what was happening in the street. He did, however, not get further than the front garden. Two Russian soldiers had caught sight of him, jumped over the fence and demanded shoes and jewellery. When Grandfather had tried to explain that we had neither, they had knocked him down and entered the house. One had pushed a torch into Mama's hand and they had forced her to hold the light for them while they searched the house. They had taken our last food rations and some of Papa's clothes and then left.

Later, when people began to meet again, we discovered that we had been extremely lucky on that first morning. During the interval in the bombardment, while we were still sleeping in the basement, the first Russians had already been roaming the streets. In the road behind our house they had raped a woman and her twelve year old daughter and shot her father who had tried to protect them. At the end of our street, three had raped a little girl of seven and she had died the same evening, 'bleeding to death', as Mama put it. They had of course taken bicycles and watches, clothes and food wherever they found any, but that was something nobody really minded. Many other things had happened in our town on that April morning but we did not learn of them until much later and by then we had got so used to rape and murder that it no longer horrified us.

In the days following that first onslaught we hardly slept. We never took our clothes off for fear it might become necessary to hide at any moment. During the night little groups of Russian soldiers were going from house to house, looking for women. They just broke in if people did not open their doors voluntarily. In most cases, people preferred to open, because once a door was broken, anybody could walk in and out without warning. It was strange how sensitive we became to sounds. The slightest noise, even streets away, was enough to wake us, and we would get up and listen to the steps and the voices coming nearer and nearer. If the moon was out we would watch them from behind the curtains, entering one house after another. Sometimes people held out longer than usual, sometimes they opened immediately. Sometimes the search party came away quickly and sometimes we could hear shouts and high-pitched screams and then we knew that it was happening.

Robert had been right. The remains of the German army had gathered in the Dunkelsteinerwald for a last, senseless stand. From there they were bombarding our town. The Russians dug trenches along the western border and shot back. Somehow we soon got used to it. After two more nights in the basement we decided to sleep upstairs. In the basement it would have been difficult to hear anybody trying to break in. Besides, we had to take the risk of being hit by shells during the daytime. Only Mrs. Donner, who had completely lost her nerve and was all the time crying, saying she would be transported to Siberia like one of her sisters had been at the partition of Poland, stayed on.

To crown matters I caught a terrible cold. Actually, it was a very peculiar cold. It did not seem so bad as long as everything was quiet, but as soon as we had to slip into our secret hiding place under the staircase, I just could not stop sneezing. Somehow the idea of having to sneeze just at the moment when I was supposed to hide seemed terribly funny and I would burst into fits of giggles with my head on Henny's shoulder.

Mama always got very angry, she would knock against the planks and tell me to keep quiet, listing all the punishments she would inflict upon me if I did not obey. Hearing the threats, which under the circumstances had absolutely no meaning, struck me as even more funny, and Henny and I had to press our handkerchiefs into our mouths as we shook with silent laughter.

I think it was on Tuesday that our new overlords issued their first order. Curtly, and in quite familiar language, we were told to deliver all arms and radios to the town hall. Those failing to do so were threatened with execution. In the afternoon of the same day Grandfather and Mama put our own, Mrs. Donner's and Mrs. Werner's radios on a little hand-cart and wheeled it to town. Henny and I were told to stay in the basement. Hans was commissioned to look out and immediately report any approaching danger, so as to give us time to hide. He felt terribly important in his role but I don't think he would really have been of much help. After some time Henny and I got bored in the dark damp basement, and since we were afraid neither of her mother nor of Mrs. Donner we just went upstairs to my room and sat down on the floor, out of sight from the neighbouring gardens.

It was a beautiful warm spring afternoon. Ever since the beginning of the bombing the good weather had held. In our neglected garden daffodils and bluebells were beginning to cover large parts of the lawn. The footpaths were overgrown with grass and nettles and patches of violets had spread from the places where Grandfather had originally planted them. Last autumn he had put potatoes at the end of the garden and they were beginning to show their first fresh leaves. To us, hiding behind the curtains of the French window that led to the balcony, it seemed the garden had never looked so beautiful.

On their return Grandfather and Mama brought the first reliable news. All the deserted flats and houses had been looted, not only by Russians but also by some local people, who suddenly claimed they had always been Communists. I heard Mama say that similar things had happened in 1918 and 1919. It was strange how such elements appeared and multiplied in times of disaster. Just like rats, that hide during the daytime. But there was something else. At the Town Hall, Grandfather and Mama had been told that all women between the ages of fifteen and forty five, who did not have small babies, had to report every morning to a place near the destroyed railway bridge over the River Traisen to help rebuild it.

Mrs. Donner immediately said: 'I won't go. I won't put a step outside this house. I am not well, I have a weak heart. My husband would be very angry if I did not look after myself properly.'

But Grandfather told her she would have to go. The police, who consisted entirely of freed ex-prisoners, would check houses and if they found her, they might make trouble for Henny and me.

Mrs. Donner began to cry. 'Why should I risk my life for those two girls? They mean nothing to me. They are young; they can stand it much better. I am not healthy; I have never worked with my hands. I tell you, I just can't do it.'

And Grandfather had to deal firmly with her.

I was rather shocked by the whole scene. After all, Mrs. Donner had always been rather nice to me, a few times she had even said she was very fond of me, and I - well - I think I had believed her. It was as if those who had agreed to look after my defence were already discussing the possibility of handing me over as a hostage.

Then began those three strange weeks, preceding the eighth of May, when the war finally ended. I don't think I shall ever live through anything quite like that. Should I suddenly be faced with a similar situation I would probably say, just like Mrs. Donner, 'I can't stand it'. But the previous months, during which life had become more and more twisted, had gradually prepared us for it. When it came, we were so dulled, we hardly knew what was happening.

The bombardment by the Germans did not stop. Neither did the Russian search parties. The whole world seemed to consist of only two groups of women: those who had already been raped and the others to whom it had not yet happened. The few doctors left in the hospital did hardly anything else but help women not to get pregnant, or catch a disease, though I overheard Mama say that if the disease was syphilis, there was nothing they could do. In the large public underground shelters, where people who had lost their homes were hiding from the bombardment, it was especially bad. There the Russians came every night and the Communist policemen accompanying them tried to persuade the women to give themselves up voluntarily, saying they should be grateful to the Red Army for having liberated them from Hitler. Actually the word 'liberation' was quite a joke in those days. People burst out laughing whenever they heard it and said, since Hitler had already liberated us from our country, and the Russians were now liberating us from everything else, we would soon be in the fortunate position of having nothing left that might attract a third liberator.

Mama and Henny went to work daily on the railway bridge. Mrs. Donner went with them twice and then simply lay down on her camp bed and refused to go. Alexandra, her uncle's servant, forged a certificate in Russian, stating that she was exempt from work. We were all disgusted and thought her very selfish.

During the daytime I was left alone with Grandfather. But even Grandfather could not stay with me all the time. Only one baker in our town had been given permission to open and so Grandfather had to spend several hours queuing at the bakery most days. We had some potatoes left from the year before, but nobody bothered much about food, somehow we did not feel hungry.

Since the arrival of the Russians the women had begun to wear dirty clothes and rub dirt and ashes into their hair and faces to make them look as ugly as possible. A young woman across the road who ventured out in a proper summer dress, paid very dearly for it. Mama spread the rumour that she had sent me into the country. People trying to curry favour with the Russians were beginning to disclose addresses of houses where young girls lived. One could no longer trust anybody.

Mama changed very much in those days. Somehow she seemed to have shaken off that strange indifference that had enveloped her since Grandmother's death. For the first time she seemed almost normal, there were none of those hysterical outbursts that had occurred from time to time before. I think she was sorry that she had been so violently opposed to Hitler. I still remembered how at the time of the *Anschluss* she had locked herself into the bedroom for days; refusing to come out, because she said she could not bear to hear people she had known all her life, suddenly greet her with 'Heil Hitler' in the street. Now Germany was losing the war and the Nazis she had hated so much were being destroyed. But what had become of our lives? Everything was exactly as Hitler had prophesied it would be should the enemy win. For once, at least, he had spoken the truth. Mama, and I think quite a number of people with her, began to feel betrayed.

Coming home in the evening, she told us how she and the other women had carried telegraph poles and loads of stones for long distances to the destroyed bridge. Some women were given the job of throwing stones from a small, swinging bridge into the river and there had been a few accidents. The Russian soldiers sat along the way with their guns in their hands, shouting orders, cursing those who were slow and pestering the young women. If somebody tried to run away, they just shot into the air. Mama also described how the German bombers attacked the bridge at regular intervals and how the women had to throw themselves to the ground since there was no shelter.

One woman had lost her nerve and cried: 'I can't stand it! I can't stand it any longer! I shall go and complain.'

And the others had laughed and said: 'To whom will you complain? The only one left is God and he doesn't seem bothered.'

The worst, however, was neither the work, nor the bombing, but the way to and from the bridge. Again and again women disappeared. Once a young officer tried to pull Henny into his car. Fortunately there was nobody about so Mama and Henny grabbed some stones and beat him over the head until he was unconscious. Afterwards Henny stayed at home with me. But that was not all; something else threatened the women on their daily way to work. All along the western border of our town the Russians were digging trenches. In the beginning they had taken men to help them, but since there were hardly any men in town, they had begun to round up women and drive them in lorries to the front. Sometimes those women were kept there for several days. They had to work at night and were forced to sleep during the daytime in the same shelter as the Russian soldiers. Mama was very afraid of being caught in this way and whenever it was time for her to come home in the evening we grew anxious. Then at the beginning of the second week it did happen. She did not come home for two days and nights. Grandfather tried to get some news but nobody knew what had happened. We had no idea whether she had been taken to the front, carried away by some Russians, or whether she was already dead. At last, on the third day, at five o'clock in the morning Mama returned. She was covered with mud and I had never seen her look so ill and so old. She just went upstairs and threw herself on the bed without taking her clothes off. Grandfather stood beside the bed, watching her anxiously. Once she turned her head to him and said: 'Well - if Hitler comes back, I shall be the first to join the Party.'

Mama did not get up for two days. She was completely exhausted and since we had nothing to eat except a few potatoes it was difficult for her to recover. Grandfather visited all the people we knew and even searched through empty houses to find something for her, but without much success. Since there was nothing else I could do, I would sit on her bed. She would talk to me a little, relapse into silence for many hours, and them talk again. Sometimes I felt almost embarrassed by the way she spoke to me. Mama and I had never been really close. I had always regarded her as someone who just inflicted punishment and spoiled whatever fun I had. To be quite honest, sometimes I had actively disliked her. She was quite different from Papa, who seemed just mildly amused whenever I got myself into trouble and, most of the time, paid little attention to me. In a way, I had been afraid to talk to her because she either corrected my diction or found fault with the construction of my sentences. With the war on, and Papa away nearly all the time, she had not seemed very interested in me, especially during the last few years when she had to do war work. After Grandmother's death her interest in me had become less. Now, for the first time, she really talked to me, not as one talks to a child to whom everything is told with explanations, but to another grown-up.

Mama's story was simple, we had of course heard it before but if things happen to you they are always different. On her way to the bridge she had met a group of women talking to a policeman. He had tried to persuade them to accompany him to the railway station. The Russians wanted it to be cleaned, he had said. The job would be very much easier than rebuilding the bridge. At first the women were doubtful. It was possible that the policeman spoke the truth, but then he might just as well have been a Russian agent, trying to trap them. By chance Mama knew the man. Years ago he had been a dustman and collected rubbish from our road. Mama had been pleased to see a familiar face and when the other women realized she knew the policeman, they had agreed to go to the railway station. Mama had joined them. There was a Russian army lorry standing by. The policeman asked them to get into the lorry. Again the women had hesitated. Stepping into a Russian car was considered an unpleasant form of suicide in those days. But in the end the policeman had convinced them that they had nothing to fear, and he had promised to sit in front with the driver. This, it was later discovered, he had somehow cleverly avoided at the last moment. As soon as the women were in the lorry they understood that they had been cheated. The Russian driver did not drive towards the station; he took the next turning to the left and drove straight westwards out of the town.

The women had protested and cried and tried to jump out of the lorry but without success. The lorry drove to a little empty village, several miles away from the town, and there they were taken into custody by a group of armed soldiers, who escorted them over the fields to a lonely farmhouse. Like the village, the farmhouse had been looted and its inhabitants were just leaving. They were an old couple with two small boys. They had no possession except a little goat, which the woman carried in her arms. When the two groups passed each other, the woman with the little white goat began to cry and made the sign of the cross.

Mama and the other women were led into a half-empty barn and the door was locked behind them. They spent a long unhappy day in the barn. Some of the women cried, and some quarrelled, and some said it was all Mama's fault. She had told them to go, they said, though, actually, she had done nothing of the kind. At lunch time, two soldiers came and said they wanted a few young girls to help them peel potatoes. Nobody was willing to go. At last three girls were simply pulled up and taken away and when they returned a few hours later, they were all crying.

Somehow, I could picture it all quite well. The large dark barn on the border of the forest, the sound of shells hitting the town and the irregular rattle of the machine guns. A short distance away from them were the last remains of the defeated German army, their own people. But they were no longer their own people. They were just a long, deadly line of guns, and in the night, when the Russians would drive them out to dig trenches they would shoot at them. Death in front and the enemy behind, there wasn't much to choose either way.

Mama stayed there for two days. In the evening of the second day she collapsed in the middle of the trench they were digging. They had pulled her up and one soldier had made her drink some vodka. But since Mama had hardly eaten anything for two days, she immediately vomited. At last they realized she was no longer of any use to them and the officer in charge sat down and wrote a certificate, saying she was exempt from work for two days. Then they told her to go home. It was in the middle of the night and the gunfire from the German side rang out in a continuous low rhythm. Mama knew her way home. We had often gone to this particular area on Sunday afternoons. But in the darkness, and with the ruined farmhouses, it was difficult to recognize the way. There were Russian soldiers all over the place. Several times Mama was stopped and had to show her certificate. They shone torches into her face but fortunately Mama looked very exhausted and ugly, and she was smeared all over with mud, so they let her go.

'I don't so much mind the Russians,' Mama said, 'after all they are our enemies. They are fighting a war and they don't like us. But the man who betrayed us. I just can't understand it.'

I said, in order to comfort her: 'When Papa comes back he will have him put in jail.'

But Mama laughed and replied: 'My dear child, if Papa comes back he will be very lucky if he isn't put in jail himself.'

In a way, it was fortunate that very soon afterwards Russian soldiers confiscated a room in our house. Otherwise Mama would just have kept on lying in bed and I think that might have made matters worse. But the way things went she had to get up and look after the house and nurse Grandfather, who had gone down with a severe attack of dysentery. She also had to protect Henny, since Henny's mother was rather stupid and of no use at all.

Almost from the moment they had arrived in our town the Russians had been looking for accommodation, especially in residential areas. People, who despite the war, had kept their houses clean and in proper order, had been forced to give up their best rooms. Some were made to live in the cellar while the soldiers occupied the rest of the house. Mrs. Donner had told us the best protection against such unwelcome tenants was to make everything look as dirty and shabby as possible. 'Because,' she said, 'that's how they live at home. It won't tempt them.' We had all done our best to follow this advice and it had so far saved us. But after the first week the Russians began to look for rooms more systematically, not only for officers, but also for ordinary soldiers, and the artificially created slum in which we lived was no longer a deterrent. They started chalking signs on several houses in the road, and though the women had secretly erased them, eventually came a day when they literally besieged our place.

Henny and I were sent to our hideout in the cellar and consequently we were unable to know what was happening. We only heard knocks, hoarse shouts and occasionally Mama's voice pleading with Grandfather not to open. At last, after what seemed hours, Mama came and knocked against the planks of our hiding place. We climbed out and Mama whispered: 'Take your shoes off and follow me upstairs.'

Shivering with cold and with very uncertain feelings we followed her. Mama took us into my room and locked the door.

'They have come,' she told us unhappily. 'Three of them. It was not possible to hold out any longer. They would have broken the door and you know what that means. Grandfather thinks you should stay upstairs. We can't keep you locked up in the basement all the time. But don't wear your shoes, and sit on the bed as much as possible, they must not hear any footsteps overhead. And for heaven's sake don't talk loudly or go near the window.'

I looked at Mama. She seemed calm and there was something in that calmness that made me feel uneasy. It was the same inhuman, detached fearlessness she had shown during the air raids that had made me feel so uncomfortable. Gone was the understanding, the nearness she had allowed me to feel during the last few days. Somehow it was dreadful not to be allowed to play any part at all, to be locked away and protected like some dead thing. I said hesitantly: 'But what happens when you go to the bridge? Shall we look after Grandfather? I mean -'

Mama said: 'I won't go to the bridge any more.'

I felt relieved. It was one thing to want a share in the responsibilities and quite another thing to be given it.

I said: 'Will that be all right? You said anybody who does not go would be shot.'

It then occurred to me that Mrs. Donner and Henny were exactly in the same position and I was overwhelmed by the feeling that I was living in a house full of people who were all doomed beyond salvation.

Mama shrugged her shoulders and said: 'I don't care, if this is life in a free Austria, we might as well be dead.'

May

The first days of May brought an unexpected snowfall. Shortly after noon, big watery snowflakes, like white apple blossoms, began to float down. In less then an hour the lawn, the balcony in front of my room, the roofs of the neighbouring houses, and the fences separating the gardens, were covered with white cushions. On the footpath and wherever the wet snow touched the bare ground, it immediately melted, leaving an ugly brown stain of mud. But it stayed and collected on the branches of the large old peach tree at the bottom of the garden, bending them to the ground. By evening the burden had become so heavy that the tree split into two. I watched it from behind the curtains and it made me feel very sad. The peach tree had been my special friend as long as I could remember. I had hidden in its branches whenever I had committed an offence that was sure to bring punishment, and, screened by its leaves, I had read all sorts of books, I wasn't supposed to read. Sometimes at night, when the full moon made me feel sad and restless, I had crept down into the garden and, shielded by it mighty stem, I would sit under it for hours. I had also made all sorts of secret promises and resolutions there, touching the bark with my hands as if to make the tree my witness. Watching the torn tree, I remembered, with some embarrassment now, how, only two years before, when I had read several books about the shameful way the white settlers had treated the American Indians, I had promised the peach tree that as soon as I was grown up, I would dedicate my life to their cause. Now the peach tree had been destroyed. Half of its branches lay in the mud, their blossoms soiled by the wet earth. A long white mark ran down the stem like an open wound. It was as if part of my childhood had gone forever. Had Henny not been with me, I might have cried.

Contrary to all expectations, the Russian soldiers in our house made us feel safer. Because now, whenever other soldiers knocked at our door during the night, our new tenants, annoyed at the disturbance, would open their windows and tell them to go away. Mama told us that three soldiers had established themselves in our living room. From one of the deserted houses in our road they had brought mattresses and blankets and bedclothes. In the daytime they busied themselves repairing shoes for their comrades. Two of them were elderly men, who were mildly drunk most of the time and took no notice of anybody. The third was a young soldier, who spoke a little German.

On the day of their arrival, he had tried to hold a conversation with Mama. He had told her his name was Sasha. He came from a little village in the Ukraine and his wife, he said with great pride, was a teacher.

Mama listened politely, determined not to give away any information. But he had insisted: 'You have husband?'

Mama said: 'Yes.'

'Where?'

Mama: 'Danzig.'

'Prisoner of war?'

Mama shrugged her shoulders.

'Prisoner of war,' Sasha said, as if he had just received some confidential information. 'You not worry. Well fed, well treated. Russian people at home - nice.'

Mama said dryly, she hoped he was right.

Sasha continued his examination: 'Children?'

Mama said: 'Yes,' and immediately regretted the answer.

'Son?'

Mama said: 'No.'

'Good!' Sasha smiled. 'If son, son fight in war, get killed, you think I did it. Good!' He repeated and then asked the question Mama had been dreading all along: 'Daughter?'

'Yes,' Mama said, adding quickly, 'but she is not here, I have sent her away.'

If the whole examination had been a trap Sasha did not show it. He smiled understandingly and said: 'Good! Girls and war, not good together,' and, after this pronouncement, he marched off in the direction of our living room.

Mrs. Donner, who had heard the whole conversation, told Mama she should not have been so friendly, but Mama said she had no wish to be friendly with any Russian.

Next day, when Mama was sitting in grandfather's room, Sasha appeared with a large saucepan, covered by a rather murky looking handkerchief.

'Papa ill,' he said. 'Must eat,' and he put the saucepan on Grandfather's bedside table.

Mama and Grandfather shook their heads and explained that they did not intend to accept his present. There were, of course, people who took food from the Russians but everybody thought them very low. Sasha looked thoughtful and then said: 'Cooked in army kitchen, not stolen.'

Grandfather said he quite believed it but we had plenty to eat ourselves. That, of course, was a lie and not a very convincing one. Sasha shook his head and looked annoyed. Then suddenly a new idea seemed to come to his mind. He folded his hands and holding himself very straight, began to recite the *Pater Noster* in almost faultless German. After he had finished he beamed and then asked: 'Good?'

Involuntarily Mama and Grandfather had to smile and they agreed that he had said it very well indeed.

Sasha looked highly pleased: 'Learned as child from grandmother, she spoke German,' he told them, 'but you not tell the others,' pointing with his thumb in the direction of our living room.

Henny and I, of course, knew this only from Mama's accounts. To avoid suspicion Mama did not go upstairs during the daytime. She came only at night, when everything was quiet, and brought us whatever food that was available and water and told us the news. She herself slept in

Grandfather's room because he was still quite ill and it was also the best place for watching the house.

There was a particular place on our landing, from where one could overlook parts of the hall without being seen from downstairs. It was not an easy thing to do though, one had to lie flat on the floor and press one's face against the bars. I had discovered this place quite by accident when I was much younger and often I had lain there, listening to the conversation of the grown ups, long after they had sent me to bed. For some unknown reason, it was possible from this point to overhear almost everything that was said downstairs. Though Mama had strictly forbidden us to leave my room, Henny and I used to sneak out and look down from the landing whenever we had an opportunity. Since Sasha and the other two soldiers were shoemakers, there was much coming and going in our house. Mama tried to spend a good deal of time in the hall in order to stop people from going upstairs. Sasha began to tease her, saying she was like a watchdog and sometimes he pretended to go upstairs and afterwards mimicked her attempts to stop him. Mama often felt he had his suspicions because once, when she had come back from queuing for bread, she found him waiting in the hall.

'Two men want to go upstairs,' he told her, 'bad men. Me, I stop them.'

And when Mama looked at him with horror, unable to speak, he added with great dignity: 'Not all Russians bad.'

There were of course unpleasant incidents too. Like the days when two officers molested Mama so badly that she had to jump through one of the kitchen windows. And the afternoon when a completely drunken soldier smashed all our China and glasses against the wall and afterwards used the big mirror in the hall for shooting practice. The most unpleasant incident, however, was the one with the Polish woman.

One day Henny and I heard some strange noises coming from downstairs and when we listened at the door it sounded like a woman screaming. For a moment we thought it was Mama and, carefully avoiding to make any noise, we unlocked the door and tiptoed out to the listening place on the landing. The woman who was screaming had not been Mama. Mama stood with her back against the wall, looking very angry and also a little frightened. In front of her stood a woman and two Russian soldiers. The woman had a high pitched wailing voice and, though she spoke very good German, she was obviously Polish.

She was saying: ' ... and not even proper tea. It is some awful herb tea you have given me. Who knows what sort of herbs they are. They might poison me.'

Mama said, trying to control her anger: 'We don't have anything else.'

But the woman paid no attention to her. She continued in the same wailing voice: 'And you have put no sugar in it. How dare you give me tea without sugar? I have asked you for a cup of tea because I am thirsty and I feel sick and you give me this bowl of dirty hot water without even sugar in it.'

Mama said: 'Stop shouting. I can't give you any tea and I can't give you any sugar. We don't have such things ourselves. Whatever we had has been stolen by the Russians.'

The woman screamed: 'How dare you speak like this? You filthy German bitch! You dare accuse Russian soldiers of stealing? You still think you can push people around. Your soldiers have been beating us and killing us.'

I could see that Mama was now in a terrible temper herself. She shouted: 'I hope they have! I shall be sorry for the rest of my life if they haven't', and snatching the cup from the woman's hand she threw the liquid into her face.

For a moment there was silence. Then the woman let out a shrill scream and, turning to one of the soldiers, she cried hysterically: 'Shoot her! Arrest her! She is a Nazi! Have you not heard it? She has just said so herself!' followed by a long stream of words in Russian.

The other soldier got hold of Mama and began to shake her. I closed my eyes; I wished Mama had controlled herself. It took some time for me to realize that the woman had suddenly stopped screaming and that somebody else was now talking in Russian. I looked down. A very young soldier was standing in front of Mama. He had fair, cropped hair and a very broad face. From all Mama had told us, it could only be Sasha. He kept on talking to the woman and the two men in what did not seem a very friendly way. Eventually the two men grinned, turned round and left with the woman.

Sasha touched Mama's shoulder and said: 'Not worry, Mamushka. She very, very bad woman. Sleeps with soldiers, steals their money. Done me once. She,' he shook his head in disgust and added, 'she whore.'

The best thing that happened in those days was the close friendship that sprang up between Henny and me. It was the first time I had ever made friends with an older girl. I usually did not like them. The older girls with their smart haircuts, their better figures, and their experience with boy friends, always made me feel childish and inferior. Of course, Anna was experienced too, but she was of the same age. I could always bluff her and tell lies to her, but somehow I had the feeling an older girl would have understood that my stories were just invention. It would have been terrible to be found out and laughed at; I just could not even visualize such a humiliation. But I must say Henny never showed off. She just told me about herself and her engagement to Robert, as if I was perfectly capable of understanding. Not like Anna, who always considered it necessary, especially in front of Sophie, to give comments and explanations. Henny told me how she had met Robert at a holiday camp and how different he was from all the other boys she had so far known. Of course, her mother would not allow them to get married before she was twenty-one. It had already been very difficult to make her agree to the engagement. Henny and I talked it over and decided it was probably just selfishness on the part of Mrs. Werner, she simply did not want to lose Henny's help in the shop. But since Henny's father had been killed in Africa two years previously, we also agreed that this was understandable. Robert wanted to become a gymnastics teacher and hoped to have his first job in four years time. In exchange for her confidence I told Henny that I wanted to be a writer and go to university and become very famous one day, and Henny did not smile the way Anna always did whenever I said something like this. She just said that was marvellous and she had always wished she was as clever as I was. It was the first time somebody about my own age had told me I was clever and I liked it very much.

In this way the days passed. We spent them sitting on my large divan bed, talking to each other in low confidential whispers. Most of the time we were very hungry and sometimes, when Mama had gone out and the three Russian soldiers were away from the house, we sneaked downstairs and looked for food but we seldom found anything. As time passed, the explosions from the German shells became less frequent, the sounds of machine guns died down, and then came silence and after it, the eighth of May. Somewhere in Germany somebody had signed a peace treaty. The war was over, Hitler had killed himself and at the last moment married a woman called Eva Braun. The Empire of a Thousand Years, about which we had heard so much at school, was over. Papa lost in Danzig, Robert in the Dunkelsteinerwald, all the people killed and wounded, all the misery of the air-raids had been for nothing. The miracle the German radio had promised us had not happened.

In a way the eighth of May was disappointingly similar to any other day. Even Mama said that in all the years she had been waiting for peace, she had never thought it would be like this. Grandfather was a little better and got up for the first time. He insisted that Mama should stay upstairs with us and get some rest; he would look after the house. In the afternoon it became obvious that the Russians planned to hold peace celebrations, and that one of them was going to take place in our garden. Bottles and bottles of Vodka were stored on our verandah and a group of Russians invaded our kitchen and began to cook. Sasha, rushing in and out of the house, embraced Grandfather, kissed him on both cheeks and cried: 'Peace, Papushka, peace! Great feasts! All friends now.'

We watched the preparations with growing apprehension. A large number of drunken Russians in our garden wasn't exactly something to look forward to. It was decided Henny and I should go into the attic. Mama would lock the door behind us and join Grandfather. A solid metal plate backed the door to the attic and it was almost impossible to break it down. It only had one disadvantage: it could not be opened from the inside. But, since the bombardment had now stopped, this did not really matter. So about four, when it became apparent that the celebrations would start at any moment, Henny and I slipped upstairs to the attic.

For some time we sat talking on the old chest. We discussed what would happen next, when the Russians would leave Austria, whether the shops would soon open and how long it would take before one would again be able to buy food and clothes and shoes and other things.

'Robert is probably a prisoner of war,' Henny said, 'but now that a peace treaty has been signed, it can't be very long before he is free again.'

It was the first time in days that she talked about the future. I had often wondered what she was feeling. Was she worried? Did she believe Robert was still alive? Used to Mama's bitter complaints and fits of despair, I could not quite understand her silence. But I had never mentioned the subject myself because I felt it was something rather private, and if she did not want to talk about it, I should respect her silence. Even now I felt a little shy and after a quick look at her face I murmured: 'Sure Henny.' And: 'Of course he will be free.'

And when she did not say anything further I changed the subject.

The noise in our garden had suddenly died down and we could hear a single voice speaking in Russian.

I said: 'Henny, if we put those two suitcases on top of each other we can look out of the roof window.'

Henny was doubtful. 'I don't know, won't they be able to see us?'

But I said: 'Of course not. Somebody is giving a speech, they will all look at him.'

The suitcases, which belonged to Mrs. Donner, were full and rather heavy but after some difficulties we succeeded. Carefully balancing our weight, we climbed on top of them supporting ourselves on the rough woodwork of the window frame.

It was just around dusk. The sun had set behind the long row of houses, drawing a few far reaching purple lines across the blue sky. There seemed to be at least a hundred soldiers in our garden. One man, obviously an officer, was standing on an upturned box in the middle of our muddy lawn. From time to time the soldiers cheered, raising their right arms with clenched fists. It was a strangely unfamiliar gesture and for some unknown reason it made me feel afraid. When at last the officer stopped, they all began to embrace and kiss each other, and some took out large red handkerchiefs and, to our immense amusement, we realized that most of them were crying.

'Well,' Henny said, 'they can't be all bad if they cry.'

Just then Grandfather appeared on the scene and one of the soldiers rushed up to him and embraced him and a few others followed his example. The two suitcases began to give way and we both jumped down at the same time and burst out laughing. It was of course too early to say but peace did seem slightly better than war.

By the time Mama came to fetch us, it was already dark. She struggled up the steep narrow staircase by the uncertain light of a candle and for the first time in weeks, called in a normal voice:

'It's all right now. You can come down.'

I asked her what had happened to Grandfather in the garden and she asked sharply:

'Did you look out of the window?'

I said:

'Well, yes but -'

Surprisingly she did not get angry. She laughed and said: 'It was that fool Sasha. He was, of course, completely drunk, but suddenly he got all emotional and wanted Grandfather to bless him and tell him he was really a very good person and a few others followed his example.'

We all laughed. The idea of Grandfather standing in the middle of our garden, blessing Russians seemed quite hilarious.

Henny's mother and Mrs. Donner were already waiting for us in my room. Mrs. Werner embraced Henny and called her 'my dear daughter' but I knew she did not really mean it. She had always preferred Hans to Henny and once I had heard her say that she had 'never really wanted the first child.' She told Henny that the Russians had stolen all their clothes, that the baby gave her a lot of trouble, and that it was high time everything became normal again. (By this she meant, of course, that Henny would do all the work for her). Mrs. Donner seemed to have regained some of her original composure and even made a few discreet remarks about her and her husband's great wealth back in Poland, a subject she had not touched upon during the last few weeks. We all talked and laughed and, after some time, Grandfather came upstairs and said we should go to bed and try to get some rest. Mrs. Werner asked Henny to come with her and the two departed through the tunnel in our basement.

Mama decided to sleep upstairs and Grandfather advised us to use the double bed in my parent's room. For the first time in many weeks we took our clothes off and lay down properly. It was quite a happy evening. Mama was in one of her rare good moods and the way she talked made me feel close to her. I always liked her when she was like this, it reminded me of days in my early childhood, when there had been less quarrels between Mama and Papa about Papa joining the Party and other things. Lying between the sheets, we discussed the near future. The worst was certainly over. Now that the war had ended the Russians would probably be less violent. Mama said that sooner or later Austria would have her own government and if Papa was a prisoner of war, it would not be very long before he was released. Talking happily we fell asleep.

I don't know what woke me. It was probably only a faint sound that had penetrated my dreams, but suddenly I was wide awake. The moonlight fell over the bedspread in front of me and when I turned my head I saw Mama standing at the window. Quickly I jumped out of bed and whispered: 'What's the matter?'

'I don't know', she said in the same low voice, 'but I heard some shouting and when I looked out, I saw six Russians go into Mrs. Werner's house.'

I asked in disbelief: 'But how could they get in? Did somebody open the door?'

And Mama replied: 'I don't really know.'

We waited. It was very quiet outside. The trees with their halos of blossoms looked white and ghostlike in the moonlight. There was only a small strip of land between the two houses but without opening the window it was impossible to hear anything.

Then I remembered.

'Henny,' I whispered, and Mama said: 'Yes, exactly.'

Again we waited. Suddenly one of the windows lit up, throwing a bright stream of light over the dark lawn. For a few seconds the silence held and was then shattered by a shrill, piercing scream.

I tore back the curtain and opened the window. The screams became louder, clearer and I recognized the voice.

'It's Henny, Mama,' I cried, 'it's Henny. The Russians have got her. We must help her, quick.'

I ran to the door but before I could turn the key Mama had reached me.

She said: 'Are you mad? There are six men in that house. Nobody can help her now.'

In the cold moon light Mama's face looked hard and cruel. It showed the same expression of indifference as it had on the day when the first bombs fell. I could not understand it and at that moment I hated her.

'You are mean,' I shouted. 'You are a filthy coward. You only think about yourself.'

Noiselessly and efficiently Mama slapped me. I tried to push her away but she kept on slapping my face and I backed against the wall. The moment of hate and rebellion was over. I saw her again as I had often seen her as a child: grown up, powerful and frightening.

'Keep quiet!' she hissed, 'and don't move out of this room. Do you think it will help her if they get hold of you too?'

I stopped crying and we both listened. The screams were still there but they had lost their touch of familiarity. It was no longer Henny's voice; it was hardly a human voice at all. I threw myself on the bed and pressing both hands against my ears. I bit into the pillow.

After some time Mama left the room but I did not move. Desperately I tried to check my imagination. I did not really know how exactly it was when one was forced to have sex with a man, I was not even sure I knew how it was when one had sex without being forced. But as much as I tried to avoid it, pictures came in front of my eyes, sick, humiliating, perverted pictures. Things I had heard at school, words I had picked up from the conversation of the grown-ups, phrases I had read in books: they all came alive, forming a pattern of revolting clearness.

Next morning Mama told me that Henny was dead. She did not tell me anything else but when she talked to Mrs. Donner in her bedroom, I slipped out of my room and listened. The door was closed so I could make out only a few words here and there. ' ... terrible ... if there really is a God ... all six of them ... she was full of blood ... they even used her through her mouth ...'

I went back into my room and sat on the bed. Only twenty four hours before ago Henny had been sitting beside me. She had talked about Robert and what would happen when he came back. And now? She was dead, yes. But what did that mean? How could one be dead when only a few hours ago one had been full of life, able to walk, speak, touch things, smell and eat? And those awful piercing screams last night - had that been dying?

The whole day I stayed in my room. I watched Mama going in and out, a Mama with a hard bitter face, who refused to discuss anything with me. Once I heard Mrs. Werner's voice from downstairs and, from the window, I could see Hans standing in the middle of our garden, kicking a large stone with his foot. I tried to talk to Grandfather but he just said: 'The less you think about it, the better. Later, when you are older, you will be able to understand. But now you must just try to forget.'

As time passed, it became more and more clear to me that I had to go and see Henny. I just could not leave it like that. I could not let her be carried away and put into a grave without looking at her for one last time. For three weeks we had been friends. More than friends, we had almost been like - well - sisters.

My chance came next morning. Through the window I could see Mrs. Werner and Hans walking up our garden path, which meant that their house was now empty except for the two old aunts. I knew they had a Russian officer living in one room but I would just have to take that risk. Carefully avoiding any noise I went down into the basement. The way through the tunnel was more difficult than ever. At the other end somebody had put a suitcase near the entrance and, since I did not know about it, I knocked my head against it in the darkness. It fell over with a loud crash and for several minutes I remained rigidly quiet, waiting for somebody to open the door and call down. But nothing happened. Slowly I crept through the basement and, feeling my way along the wall, I went up the stairs. The door gave a loud creaking sound but the incident with the suitcase had made me bold and I slipped through it quickly. Standing in the Hall I had no idea where Henny was.

It was sheer chance that the first door I opened was the right one. A large wooden box, supported on two chairs, stood in the middle of the room. The dining table had been pushed back against the wall and two chairs stood on either side of the box, bearing half burnt candles and some faded flowers. A strong cold fragrance filled the room. A smell of decaying flowers, stale air and burnt wax.

The box was a make-shift coffin and Henny was lying in it. She wore a white dress - I thought I recognized it. It was the same she had worn at her confirmation. They had folded her hands and wound a rosary round her fingers. The curtains were drawn untidily in front of the windows and a bright shaft of sunlight fall across the room and over the dead face. Slowly I went nearer. It was a pale yellow face, full of bruises and there was a broad colourless scar across the forehead that made it look like the cracked head of a doll. There was something wrong with the mouth. One corner hung down, limp and distorted, as if the lips had been torn; '...even through her mouth...,' Mama's voice rang out clearly and distinctly as if she were standing beside me. For a moment I felt a nauseating desire to vomit but then it passed and nothing was left except hate and anger.

I hated everybody. Mrs. Donner and her meanness, Anna with her stupid showing off about boy friends, the people who had killed Henny, the man who had torn her mouth. I hated the fathers and mothers who lay down in bed and produced children and could not protect them. The grown-ups with their lies and pretensions. The people who had promised us an empire that would last a thousand years, and the others, who claimed that eternal bliss awaited the virtuous. I hated the men and the women and the children and God, who had created the world and was not moved by its miseries.

Most of all I hated that fabulous thing - sex - about which they made so much fuss. They kept it hidden from us and they warned us about it, talking in whispers, at the same time associating it with love and happiness. If it could lead to a beastly death like this, the best warning would have been the truth.

I stretched out my hand and touched Henny's fingers. They were cold and stiff. They were no longer the fingers of a girl I had known but hostile, unfamiliar things from another world. Suddenly I knew that God was a fiction. Morals had been invented to cheat people. I would never again believe anyone. I would never love anyone. Whenever a man kissed me I would remember Henny's face and that, surely, would not allow me to feel any unnecessary emotions. I would be free. I would just use people. I would take whatever pleasure I could and not care for anyone. After all, life was short; that bloody bastard that would start the next war might already be born.

Aloud I said: 'There is no God, and if there is, he can go to hell!'

In the evening Mama told me that she would have to leave very early next morning to go to Henny's funeral.

I said: 'I am coming too.'

Mama gave me one of those stern looks that in the past had always cowed me into obedience and said: 'No!'

I said: 'I shall come, you can't stop me.'

Mama said: 'Now listen - '

I cried: 'No, I won't listen. You have no right to stop me; you have no right to tell me anything. Where were you when Henny was killed? You didn't help her. Her mother didn't help her. Nobody did anything for her. Just because everything is safe again, you think you can order me around, treat me like a child, I am not a child!'

And then I said: 'If you slap my face, I'll hit you!'

I expected Mama to be absolutely furious. But she just looked at me and then she said: 'All right you can come. Stop shouting.'

It was the first time Mama had ever given in to me. I felt exhausted and sick and strangely frustrated.

I shall always remember Henny's funeral. It has become a hallmark by which I divide time into two halves: before and after. Before: that was the time of childhood, dim and uncertain, full of details that are difficult to connect. It was also a period of roundness, of harmony, covered by a soft shine like that of old china. After: that is an ever growing merciless clearness, a chain of sharp details and a long crack across the surface. As if suddenly a cold light had fallen on everything and for the first time I could see things as they really were.

The cemetery was on the other side of our town and since the transport system had broken down long ago, we had to walk. It was just as it had been when we used to visit Aunt Paula's grave, but with the disturbance of the last months it took much longer. Mama had rubbed dirt and ashes into my face and my hair and given me a shapeless old dress to wear. In one corner of our garden I found a few tulips and despite Mama's protest I took them with me.

We were altogether six people: Mrs. Werner, Mama, a sister of Mrs. Werner's husband and two women from our road. Grandfather was still too weak for such a long walk and Hans had been left behind to look after the baby. I had not been out in the street for weeks and our little town looked altogether different. It had changed during the last months. It looked shabby and battered, and there was a general air of neglect about everything. The gardens were all overgrown with grass and nettles, the fences mostly broken, and some of the houses I knew so well from my daily walk to school, had been heavily damaged. In one place a whole street corner had disappeared. Somewhere else, only one room had been hit on the top floor and the door to the landing was still hanging on its hinges, giving the whole scene the unreality of a theatre set. When we reached the centre of the town, we saw that nearly all the large shop windows were broken, litter lay in the streets, and the air smelt of brick, dust and rubble - and something else. There were Russians everywhere. They wore brown dirty-looking trousers and funny shirts that looked like blouses. Some of them had a red band around one arm, with some Russian letters printed on it, and they carried guns. Looking at them from the corner of my eyes, it seemed to me they all had the same faces: broad flat noses, high cheekbones and slanting eyes. I walked between Mama and Henny's aunt and Mama kept on telling me not to look up.

At last we reached the old market place, which in my childhood had been called Rathaus Platz. After 1938 it had changed its name to Adolf Hitler Platz and when we looked at one of the house walls we read Stalin Platz. In front of the Town Hall was a long pole with a red flag and three tanks were standing in centre of the square. The place was full of Russians, all fully armed and Mama said: 'Let's go by way of the promenade.'

We followed her advice and slipped back into a side street.

The road to the cemetery crossed the railway line. But when we reached what had been the thoroughfare underneath, we saw that the place had been heavily bombed. There was no sign of the tunnel, only heaps of muddy earth, damaged railway carriages and a sad forest of rusting rails sticking into the air. With great difficulty, we began to climb over the rubble and were immediately stopped by a Russian patrol. They shouted at us and waved their guns, driving us back.

Mrs. Werner began to cry.

'My poor daughter,' she sobbed, 'my poor daughter. I can't even go to your funeral.'

The women were trying to explain that they were going to the cemetery and when this did not produce any effect Mama pointed at Mrs. Werner and said: 'Funeral. Her child. Killed by Russian soldiers.'

Everybody gasped in horror. The two soldiers looked at each other and the one of them said: 'Karasho, go - tawai.'

We continued.

Once there had been a long slowly rising road, bordered by open fields, on the other side of the railway line. I remembered it very well. On top of the hill it ran past the old inn, which stood just before the entrance to the cemetery. Not even a trace of the road was left. At first we tried to follow the railway line but then we realized that the vast field of bomb craters in front of us could only be the old road. A small footpath meandered between the ditches, still wet and muddy from the rain that had fallen the night before. We had to go in a single line, one behind the other, and several times we slipped. Nothing indicated that the path went uphill and after some time we began to wonder whether we had lost our way. But then suddenly a half burnt out ruin appeared on our left: the remains of the old inn.

The priest was already waiting for us in the little cemetery chapel. The make-shift coffin stood in front of the altar. Someone had put a crude lid over it. There were no flowers and no candles. We knelt down on the cold stone floor and the priest said his blessings. He said something about God knowing best, which to me seemed an outright indecent remark.

Two old men put the coffin on a little handcart and pulled it out of the chapel. We followed, behind the priest.

The rain had started again and there were pools of water everywhere. Some of the bombs had hit the graveyard and several tombstones lay broken and scattered across the footpath. Here and there, we saw deep holes in the ground and a sweet, repulsive smell seemed to linger around some places. The graves looked neglected, grass, weeds and big nettles were growing everywhere. On the path, between the tombstones, on the graves, making them almost undistinguishable from their surroundings.

Mrs. Werner was sobbing bitterly but I was not impressed. She had never been very nice to Henny when Henny had been alive and to show her affection now seemed sheer hypocrisy. Mama did not cry but the two other women from our road did, and I did not think much about this either. I had heard them once make some very nasty remarks about Henny and her friendship with Robert.

The two old men had great difficulty lowering the coffin. At one point they began to quarrel with each other, and in the end the rope broke and the coffin crashed into the open grave. The women bent their knees and I threw my tulips into Henny's grave. I did not want to kneel. For what reason? And in front of whom? Henny was dead and if the way she had died meant divine justice, I had no wish to bow to it.

When we went back between the rows of broken graves we met an almost identical little group coming from the opposite direction. The woman in the centre did not cry but her face showed such pain and bewilderment that for the first time I felt a lump in my throat. When we came nearer we saw that it was Mrs. Schultz. We were surprised because we had not known that somebody from her family had died, so we stopped and the women exchanged a few words of sympathy. Nobody wanted to ask any direct questions but Mrs. Schultz told us in a flat voice that her husband had been shot, when he had tried to protect a niece, who had been staying with them.

I still don't know whether Mr. Schultz had been a Communist or a Socialist. But I don't think it really matters. In an old history book I had read that it is irrelevant how a man lives, the only thing that truly shows his worth is the way he dies. Mr. Schultz had died like a hero.

Two days later I had to go to bed with high fever. There were, of course, no doctors who could have told us what was wrong with me, but Grandfather and Mama thought it was something like typhus

June

It took almost a month before I was well again. It was an unpleasant, humiliating disease, with fits of high fever, heavy dysentery, hours when one thought the worst was over, followed by violent attacks, frustrating relapses. I was afraid to eat anything because, however little it was, it immediately brought on another attack. In the second week my hair began to fall out. Whenever Mama combed it, she pulled out whole handfuls, and my two plaits grew thinner and thinner. I was afraid I might become bald, and what frightened me even more was the fact, that Mama quite openly agreed that this was indeed a possibility.

Sometimes, lying there exhausted by fever and pain, I wondered whether God did exist after all and was now punishing me for having defied him. I wondered whether I would die and in my dreams I saw the nuns, who had taught us during our first years at school. They stood around me in a large circle like big black birds, their dresses and veils fluttering in the wind, their white hoods shining. They raised their fingers like big hooks, and in a low chorus, they chanted all those terrible stories about hell fire and the torture of the damned that had frightened us so much as children. Sometimes when I awoke from such a dream, I was for a moment so paralyzed by fear that I could not even move a finger. But, despite everything, I did not retract my words. I did not want to humiliate myself by asking a non-existing God for mercy. Even if he did exist, I did not want to have anything to do with him. If he had the power to punish me, he could go ahead, but I would never forgive him for the way Henny had died. There were moments when I felt terribly alone, alone in a new and hopeless way, and there were moments when I felt simply wonderful, free, strong and important.

Several things happened during my illness. Sasha and the two other soldiers had left soon after the end of the war. Sasha had told Grandfather in strictest confidence that they would all be sent to Siberia for some time to be 're-educated'. But, of course, we did not believe them. Mrs. Donner and her relations were ordered to go back to Poland from where they had been 'abducted by the Germans'; Mama told me that Mrs. Donner had been absolutely frantic. She had said again and again she knew very well what that meant. They would never reach Poland. A camp in Siberia was the most likely destination. Mama and Grandfather did not really like her anymore. In times of danger she had been extremely mean and selfish. But, seeing her in such distress, Grandfather said that if, after arriving in Vienna, she found her suspicions justified, she could always come back and stay with us. The beautiful young wife of Mrs. Donner's uncle had died two days before the transport was to leave our town, and the old man, no longer burdened with an invalid, had disappeared overnight with his little daughter and the Ukrainian servant woman. There were rumours that he was trying to reach the American Sector behind the River Enns, and that the servant woman was in reality his mistress, but I could not really believe it. Mrs. Donner was terribly upset and said he had betrayed her. Why hadn't he taken her with them? She went away with the rest of her relations but three days later appeared in the middle of the night. In Vienna, she said, she had been quite sure that the train was meant to go straight to Russia and she had left all her luggage and all her relations behind and slipped away at the last moment. She stayed with us for six more months until her husband, who had been made a prisoner of war by the Americans, came and took her to the Tirol. After that we never heard from her again,

A week after the three soldiers had left we had to take in a Russian major. He did not want to sleep on the floor like Sasha and the others, so Mama had to arrange a proper bed. He also expected Mama to cook his food and whenever he brought a Russian woman to sleep with him, he asked Mama to change and wash the bed sheets before and afterwards, He was drunk most of the time and when he gave a party, the noise went on until the morning. But he never molested Mama and he took no interest in our house. Mama used to feed us with the food she stole while cooking. Since Grandfather was too old to work, she herself had to cook for the major (how I am not quite sure) and could not leave the house and, since Mrs. Donner simply would not work, nobody in our house had a ration book, and without a ration book one could not even get bread.

When my condition became worse, somebody told Mama that Vodka was a good remedy, so Mama began to steal vodka too. This was quite dangerous and she had to be very careful to refill the major's bottle with water, always hoping that in the evening he would be too drunk to notice the difference. (An odd change had come over Mama; she suddenly seemed to be able to manage. Unfortunately, after the end of the war she began to retreat into herself again and I sometimes wondered whether I had not imagined much of what I thought had been her change during that period.)

At last I could get up. June was almost over and the weather was still good. Everybody expected a long and hot summer. The trouble was we did not really need a long and hot summer. One could not sit in the garden - the Russians in the surrounding houses might have found it provocative, one could not go for walks in the countryside - there were still cases of rape in lonely places. Even swimming was impossible. The Russians used the local pools and it would have been madness to sit among them with practically no clothes on. Besides, Mama kept telling me, most of them suffered from syphilis and it was very contagious. In those days people talked a lot about this particular disease and I knew that some of the women who had been raped had caught it and could not be cured. The doctors, who had returned to the town and now worked in the hospital, did not have much medicine in any case. Mama did not want to talk about this particular subject, but the few indications she made frightened me even more. In the end I was not quite sure how one actually got it. I was terrified to touch anything downstairs, where the Russians had been, lest there might be germs.

At last my hair stopped falling out, but I had become rather thin and Mama allowed me to cut it to shoulder length. On the day I got up, I sat for a long time in front of the mirror. I found myself looking different - more grown up and, since I had lost a lot of weight, taller. Studying my face for the first time consciously, I told myself, that the long period of childhood, which had moved slowly like a wide river in a vast plain, was over, and from now on life would move with increasing momentum. Whatever it meant to be a woman, was no longer a thing carefully locked away in the next room, it was only a question of time before I would experience it.

Anna came two days later. While her mother stayed downstairs, Anna came and sat with me on the balcony. Mama had put blankets all around the railing so that nobody could see us. Immediately Anna began to relate her adventures. Having had full faith in the English radio, which had described the Russians as 'liberators', her parents had refused to take precautions. When the first Russian soldiers came to their door on that fateful April morning, her parents had opened the door to welcome them. One of the soldiers had immediately pushed them against the wall with his gun and another had searched them for watches and jewellery. They had ransacked the house. When they had found Anna upstairs, they had torn open her blouse and taken her necklace.

'It was dreadful, dreadful!' Anna said, raising her hands in a theatrical and most effective gesture, 'Of course, mother would not let me stay at home after that. There was no proper place for me to hide except in the cellar and I couldn't possible stay in the cellar. It's so damp there. You know how easily I get into trouble with my throat. Fortunately, my father knows the Secretary of the Bishop and he went to him and asked for help. And the Secretary allowed my father to take me to the Bishop's palace, where they kept me hidden from then on. Everybody was very nice to me and they said it was simply marvellous how brave I was. Of course, outsiders are not usually allowed to stay in the Bishop's palace but the Secretary of the Bishop is an old friend of my father's ...' I was not surprised. Anna's parents would always try to get advantages for themselves, whatever the situation.

I interrupted her impatiently: 'Well, then everything was all right. You can't really say you had a dreadful time. You were quite safe. Think of what happened to poor Henny and all the others ...'

But Anna showed no interest in Henny. She looked me over in that particular way, which in the past had always made me feel uncomfortable and a little inferior, and exclaimed: 'Good heavens, you have got thin.'

That really annoyed me. So far Anna had always made fun of me because I was fat and now she made the word 'thin' sound like an offence.

'Yes,' I said coolly, 'I believe my figure is now much better than yours.'

For a moment she was speechless. Anna could be very nasty in a sweet sort of way and usually she got away with it because she was so quick that most girls did not know what to say in return. This was the first time somebody had really hit back.

However she quickly recovered and asked with false sympathy: 'You had typhus. You must be very careful, you know. It can have quite a lot of unpleasant after effects.'

I shrugged my shoulders. 'Nonsense,' I said, 'I did not have typhus. You know how Mama exaggerates everything.' (Actually Mama did not exaggerate - it was Anna's mother who did.) 'I just had a touch of dysentery; good for the complexion.'

She did not like that either. She smiled in the same way I had sometimes seen her mother smile and asked: 'Have you been raped?'

'No,' I said. 'Have you?'

'No, no, of course not,' she cried. 'What do you think? I would have killed myself. It would have spoiled all my chances of making a good marriage.'

I said: 'Why? It would not have been your fault.'

But she waved that aside.

'Don't be silly. What do men care whose fault it is? Do you know my mother heard that it had happened to Sophie? Poor Sophie, I am so sorry for her. She will never be beautiful and now that - not even her father's money will buy her a good match.'

I looked at her and for the first time I realized that she was mean. Not just selfish and catty but really mean. She had no sympathy for Henny or Sophie. What had happened to Sophie - if it had happened at all - did not touch her in the least. I suppose in a way she had always been jealous of Sophie because one day Sophie would be very rich, and she wouldn't. She now thought fate had helped to compensate.

I said nothing and continued to be sweet to her. But I knew I would never again be impressed by her showing off. I would still treat her as a friend, but I would never really like or trust her, and I would never again be quite honest with her.

July

In the first week of July came a letter from Papa. Actually it wasn't a real letter, just a rather dirty envelope with a Viennese stamp on it. It contained a small piece of paper. The paper seemed to have been torn out of an old exercise book. There were greasy spots and finger marks all over it, and Papa had only scribbled a few sentences with a pencil. 'Do not worry. I am in Moscow. I shall come back.' That was all.

For a moment we were mystified. How could Papa be in Moscow if the letter had been posted in Vienna? At first Mama thought Papa had already reached Vienna but Grandfather disagreed with her. He said Papa was obviously not allowed to write and somebody had smuggled the little note out for him. (This proved correct. It had been a Russian officer who had smuggled the note out. Papa was not to return until 1955.) For a few days Mama went around looking cheerful and more the way she had looked a long time ago, when I was still a small child. 'I shall come back,' Papa had written, surely that could only mean, he would be with us soon? But nothing happened. As time passed, Mama's happiness faded away and everything returned to normal.

Two months had now passed since the end of the war. Slowly our little town began to recover. Austria was divided by the river Enns into two (un)equal parts. Whosoever wanted to pass the line of demarcation had to obtain a special permit and even then, such a journey was far from safe. Some people tried to cross the border secretly and some succeeded. To us, people living in the American Sector seemed superior human beings who, by some miracle, had already reached paradise. They lived entirely on cakes and chocolates, drank something fabulous called Ness-café, and could go out at any time without being afraid. There were occasional rumours that the Americans might occupy the whole of Austria, but they always proved false.

Our town was ruled by a Russian Commandant and a Communist 'Burgermeister' who, though many people hated him because he was a Communist, could not really be called a bad man. The Russian High Command was in a large villa quite near the road where we lived. There were always drunken officers hanging about, who molested every woman they caught sight of. On one occasion, the guards had pulled a passing girl into the garden and raped her. People living in the surrounding houses had heard her screams and warned everybody. But nevertheless the situation was definitely improving. The Communist policemen, who had terrorized the town in the first weeks after the occupation, were slowly being replaced. When the Russians had come, they had opened all the prisons and treated all the inmates as liberated political prisoners. Quite a few were just ordinary criminals and they began to take revenge on the people, who had brought them to justice. Nobody was safe from them and a number of people, who had themselves been persecuted by the Nazis, disappeared or were tortured. A friend of Mama's lost her fifteen-year-old son in this way. In a little village to the south of our town, two such 'policemen' got hold of a lawyer and burnt him alive.

Now that the first wave of violence had passed over, people began to return to their homes. Wherever one went one could see them making repairs or pulling down ruined parts of their houses that had become dangerous. It was as if a great tidal wave had hit our town and was now slowly sinking back into the ocean. Though there was still debris everywhere, people started to clear up. Grandfather worked as bricklayer in those days. Many of our friends had their homes damaged and it was very important that the worst should be repaired before the winter came. Mama and I used to take our hand-cart and steal wood from empty bomb ruins, and in the evening Grandfather and I would cut it with an old hand saw in our garden. It was a terribly boring job but I put up with it because I liked going out with Mama. To walk down a road, openly, without dirt being smeared into one's face, seemed a triumphant confirmation of one's survival. I liked to watch the women tentatively wearing dresses made of old curtains or other strange remnants, the children playing in the neglected gardens, the wild roses that seemed to blossom everywhere, and the rich green colours of the trees. The world was full of beautiful things I had never noticed before. It was as if life itself had been lying hidden, asleep, and was now bursting into the open. Everything seemed to confirm the miracle that I was still alive, that I could see and walk and smell, and at the moment nothing else mattered. Of course we still had to be careful. It was always better to move in groups, and one never, never had to be seen after sunset, because the roads became unsafe as the evening progressed. It was sometimes a little difficult to reach home in time, because the hand-cart Mama and I used was heavy and the roads were still uneven.

The new Town Council had distributed proper ration books, but they were not of much use, and the black market was booming. Money meant nothing. The currency was cigarettes, saccharine pills and car parts. Some people became rich overnight, they moved into big empty houses and refused to leave when their rightful owners returned, with the excuse, that only Nazis had run away from the Russians. If one wanted to buy food, one had to walk for miles to some little village and offer jewellery, clothes, linen or furniture in exchange. The farmers became very arrogant and unkind. It pleased them that the town people, who had always looked down on them, were now begging for bread at their doorstep. They made fantastic demands and very often refused to sell anything. In some cases they drove people away with a whip, or set their dogs on them. Mama went back to her sewing and began to make dresses and that helped us over the worst. She used to stay away on some farm for days and often we did not know where she was. Sometimes she took me with her so that I could get something to eat, and help her carry home whatever she had been given: potatoes, a little flour, a few apples, a loaf of bread and, very occasionally, a piece of meat Some farmers were really quite nice but some were simply disgusting. They sat in their kitchen and gave us a piece of bread for lunch while they ate meat and sausages in front of our eyes. The worst part of such excursions was the journey home. We had to walk for hours over lonely country roads, carrying whatever we had managed to get. Whenever we heard a car or a voice we hid in the fields.

I was often tempted to pray at that time, I had been used to it since childhood and it seemed a normal reaction in times of danger. But I always stopped myself. At first it was difficult. I felt as if something essential was missing from my life, especially in the evening. It was not always easy going to bed without a little prayer. Often I wondered whether something dreadful would happen during the night as punishment, because, after all, I was not yet absolutely sure that God did not exist. But eventually I got used to it and as time passed I began to lose interest. It was as if I had left a safe harbour and sailed into the open sea, or cut a rope that had so far supported me. Slowly my feeling of insecurity became mixed with pride.

In any case, things were definitely taking a turn for the better. A number of shops had opened and though they had hardly anything to sell, it was nice to see them open just the same. The hotels were all occupied by Russian officers, but some hotels gave dance parties in the evenings and there were posters in the windows inviting everyone. Of course Mama did not allow me to go (I didn't know how to dance anyway) because the place was full of soldiers and black marketeers and people said that only prostitutes went there. Once, when we passed such a place in the late afternoon, I saw a few girls standing outside the entrance. They looked very pretty in their nice summer dresses, their beautiful made-up faces and long glossy hair, and to me they seemed not only cheerful but also (this may sound funny) rather kind. I said as much to Mama but she pressed her lips together and replied: 'Don't be stupid. If you ever run around in public looking like that, I shall ... ' and she stopped. I knew she had wanted to say 'slap your face', but I think she suddenly remembered our last discussion on the subject.

I did not say anything more but I would have liked to talk to some of the girls. Just to find out what they thought. And why they did it, whatever it was they did. But since Mama seldom let me go out alone, it was not possible.

Two of the cinema halls in our town had opened and they showed American and English films, also a few Russian ones but, of course, nobody wanted to see a Russian film except Communists and Russian soldiers. Before the beginning of each film, there was a short newsreel showing the terrible things the Nazis had done to other people. But they failed to impress us. Terrible things had just happened to all of us and we were short of compassion. Films about the Nuremberg Trial were even less impressive. After the way the Russians had behaved in our town, it was outright funny to see them suddenly turn moral.

At first I was only allowed to go with Mama but eventually Grandfather put in a good word and I got permission to go alone in the afternoon, providing I went home straight afterwards. Some of the films I liked immensely and I saw the 'Man in Grey' at least six times. (I was also totally fascinated by James Mason, a fascination that lasted until I went to university.) There was not much one could do in those days. Everything seemed suddenly flat. Life had become different again. The tension of the last months had gone. I felt like an onlooker in play in which nothing much happened. So I used to slip away in the afternoon, sit in the dark cinema hall and watch the love scenes. I tried to imagine how I would feel in the place of the heroine, being held, caressed and kissed by a man. But I could only imagine it up to a certain point. Sooner or later I remembered what came after kissing, and what had happened to Henny, and then I was torn between curiosity and revulsion.

I think Austria was trying to form some sort of government and there were talks about elections but I was not really very interested in it. What bothered me more was the question whether our school would open in the autumn. The Convent had now got back all its original property and one day Mama received a letter from the Mother Superior, asking if I planned to join in September. Mama was rather pleased. She had been very unhappy about the change, because it was her firm belief that the Convent school was the only place where I would not come into contact with badly brought up children, who might talk about 'all sorts of things'. By 'all sorts of things' she meant of course sex.

So for the first time in many years, Mama again rang the brightly polished copper bell at the door of the old Convent building and a smiling working sister dressed in blue opened. Everything seemed the same. There was still the statue of the Virgin Mary in one corner of the dark hall with a little lamp in front of it, and the glass door guarding the winding staircase, only to be used by nuns, that had been a continuous source of curiosity and awe for us when we had been children.

Another Working Sister showed us into the Visitors' Room, a dark circular chamber, faintly familiar. We waited for some time until, at last, a nun came in who, as her dress showed, belonged to the teaching staff of the Convent. I had never seen her before. She was tall and slim and without the black veil and the white band across the forehead, her face might have been beautiful. I wondered why she had joined the Convent. A broken romance or conviction? One of Mama's school friends had joined the convent because her fiancé had fallen in love with her sister.

I said: 'Kuess die Hand, Mater,' a formula of greeting I had not used for several years that, as I remembered, had been compulsory during my first few years at school.

She returned my greeting without a smile and after we had sat down, gave her attention to Mama: 'Had I been in the Convent School before? Yes? Oh, good!' Then we probably understood that this was a Catholic school which, apart from providing an excellent academic training, aimed at educating its pupils to proper Christian womanhood. Mama said she knew and appreciated this point. With a graceful gesture the nun threw her black veil over her shoulder and continued. Did I have any brothers and sisters? No? I was an only child? What was my father doing? Oh, he was probably a prisoner of war in Russia? The nun raised her eyebrows. Well, she hoped Mama appreciated the fact that this was a private school; there would be - fees, yes? Mama said she fully understood. The nun did not look entirely satisfied. She also wondered whether Mama knew that the Convent had suffered severe damage during the war. If the school was to run smoothly it expected, well - help from the parents of its pupils? Mama replied, without enthusiasm now, yes, she understood that too.

The nun turned her attention to me. Did I look forward to joining the school in the autumn?

Dutifully I replied: 'Yes.'

'When did you last go to confession?'

'Well, I think, a year ago.'

She looked at me sharply.

'What do you mean, you think, a year ago? Did you not go at Easter?'

I said: 'Not this Easter but the Easter before.'

The nun turned her head and gave Mama a long, reproachful look. Mama said quickly: 'I could not send her to confession at Easter. You know yourself what terrible times we went through. The bombs and the front approaching. There was hardly time for anything.'

The nun pursed her lips and said: 'There is always time to do one's duty towards God. We, in the Convent, have attended our weekly confessions without fail.'

She sighed and added: 'We are, of course, aware that we shall have to make concessions at the beginning. These poor children have been growing up in a spiritual wilderness; one must not judge them too harshly. After a year, perhaps, the worst will have been drained out of them.'

Neither Mama nor I said anything.

The nun rose and announced: 'We shall say a little prayer in church before you go. You will do well to ask God for forgiveness. It is one of the most terrible sins to miss the sacraments at Easter.'

We followed her through the long dark corridors and down the steep winding staircase, where she could walk only with difficulties because of her dress. I felt uncomfortable and angry. I saw no reason why I should ask anybody's forgiveness. The nun seemed rude and unreasonable and I already hated the thought of joining the school in the autumn. Mama had talked so much about the Convent. How cultured and kind the nuns were in comparison to the rough German teachers she did not like. The picture built up in my mind had therefore been entirely different. I had expected a few gentle words of sympathy, an assurance that whatever horrors had happened in the past were best forgotten. Instead there were cold inquiries, reproaches, and something I could only call attempts at brainwashing. I was thoroughly disillusioned.

We entered the chapel through a little side entrance. The nun took holy water from the bowl beside the door and, bending her knees, made the sign of the cross. Mama and I followed her example. Then she knelt in one of the pews and folding her slim hands and bent her head. Mama and I did the same.

It was very quiet in the little chapel. The only daylight came through a small circular window in the dome, which left the room behind us in almost complete darkness. I remembered everything quite well. During my first few years at school I had heard mass in the chapel every morning, and though I had not visited it for several years, all the details were still familiar. There was the altar with the picture of the Virgin Mary in a blue dress, flanked by golden pillars and cascades of angels on both sides. The Archangel Gabriel stood in a corner holding a burning sword, and opposite him was St. Sebastian chained to a tree, with arrows piercing his chest, a gentle mournful expression on his upturned face. In my childhood all those saints had been alive and real to me. A nun had once told us that they would speak out if one of us had an unholy thought while in church and I had often expected the Archangel Gabriel to point at me with his sword, because I was never quite sure whether my thoughts were holy enough. In those days the chapel seemed filled with a strange atmosphere, comforting and awe-inspiring at the same time. Kneeling in my pew, I tried to recall traces of those feelings but it was impossible. Now that I no longer believed that the pale host in the golden tabernacle was the body of Christ, the church was nothing but a small dark room, cold and empty.

On our way home Mama remained silent for a long time. I wondered whether she too was disappointed. In the end she said, rather harshly: 'Well, at least you will learn some manners. The sort of language and behaviour you picked up in that Nazi school were simply shocking.' Then I knew that she shared my feelings.

But we had no time to argue about it because it was already late and we had to run to reach home before sunset.

August

On the sixth day of August the Americans dropped something on a Japanese town that afterwards everybody referred to as The Bomb. I did not really understand what it was all about because, so far, we had not done much physics at school. But it was something completely new, something that had never happened before in the history of the world. The atom, which everybody so far had thought was the smallest possible thing, had been split and this had thrown out a tremendous amount of energy that was able to destroy and burn everything. Some people who had managed to hide their radio sets, listened to the news and spoke of what they had heard in whispers. Some people said they had seen pictures in foreign newspapers, secretly smuggled over the demarcation line. Pictures showing vast fields of rubble, women with dreadfully burnt faces, shadows on the wall that had once been people, disfigured babies crawling over the ground like wounded animals and pieces of shrunken and burnt flesh - the last remains of human bodies. One newspaper wrote that the deadly rays which had been created by the explosion were still hanging over Hiroshima and Nagasaki, and that for quite some time the rescuers had been unable to enter the area. There were also pictures of people who had escaped The Bomb. They looked just as dreadful as the people the Allies had liberated from German concentration camps, but somehow I think they did not receive quite as much sympathy. People did not pray for them in churches, nor did the priest denounce their murderers from the pulpit, asking God to punish them for their wickedness. I tried to discuss the question with Anna but she just shrugged her shoulders and said: 'Well, after all, they are only Japanese,' and many people said the same.

Grandfather alone was seriously upset. I watched him go around, shake his head, and sometimes he would talk to himself. I once heard him say to Mama: '... one has to be grateful for being old. God only knows what is still in store for the next generation.'

I felt only anger. Of course, I was sorry for those people. The fact that they were 'only Japanese' would hardly have made it easier for them. But at the same time I was simply too exhausted from all that had happened to be capable of much compassion. I just felt cheated and furious and impotent. There we were, with one war hardly over and those blasted grown-ups were already thinking up nice little gadgets for the next one. The world was only just becoming beautiful again and now everything seemed threatened. There were rumours that the Russians and the Americans did not really get on with each other, and that sooner or later there would be another war, a war in which they might use The Bomb. What really got me down was the fact that I was still too young for many things, and if I did not hurry up, it might soon be too late.

Towards the middle of the month Sophie and her parents returned. They had not been able to reach Mariazell and the advancing Russian army had caught up with them in a little village near the border of Lower Austria. I believe they had suffered quite badly, but since Sophie herself never talked about it, I did not ask any questions. During their absence, their house had been taken over by their former maid and her family and Sophie and her parents now lived in the basement. They had lost nearly all their belongings and, I think, felt very humiliated by the situation, especially Sophie's father.

Sophie herself had changed little. I shall always remember the first time I met her again. She wasn't at all like Anna. She did not show off about her suffering and she did not ask me whether I had been raped. She did not even remark on the fact that now they were all living in a dirty basement instead of the beautiful house that belonged to them. She just said with her lopsided smile: 'Let's sit on the bed. It's a little chilly down here.'

So we climbed on the old divan bed, which together with a shaky wardrobe and a battered looking table, were the only furniture left, and drew the blanket around our legs. For a short beautiful moment it was almost as it had been before. The basement room had just one small window and only its upper part was above ground level. I looked up and saw masses of red and yellow roses, all tangled together. The sun was shining on them from behind and made them look like transparent jewels.

I said, anxious to keep the atmosphere of happiness alive: 'Lovely roses.'

Sophie replied eagerly: 'They are nice, aren't they. Actually they look much better from here than from my old room.'

I looked at her to see whether she really meant it and when I saw she did, I really respected her.

Sophie told me about the village where they had stayed, the friendly landlady and the German soldiers who had not been German soldiers at all but Austrians from our town. They had stood behind their guns and shot and cursed and prayed and shot again because they no longer knew whom they were hitting: the Russians, their own houses, their own wives, their own children. Sophie's mother and some other women had begged them to stop, but they had said this was impossible, there was still a war on and an order had been given and they could not disobey this order and run away like cowards. Sophie's mother had slapped the face of an officer and the officer had done nothing about it. That more than anything else made me understand how bad the situation must have been because, in those days, one could not slap an officer's face without being immediately shot. Besides Sophie's mother is very sweet and gentle and I don't think she had ever hit anybody before in all her life.

And then I told her about Henny. I told her about the night of the white moon, the halo of blossoms in our garden, the screams and the silence, the dead 'thing' in the coffin, and the funeral, which the grown-ups had turned into a mockery, just as they did with everything else. Sophie said nothing, absolutely nothing. I knew then that though she could never be my friend the way Anna had been (you have such a friend only once) I would from now on always love her best.

Our school was supposed to start in September and, since we had done hardly any studying since the beginning of the year, Sophie's father suggested we should work together for a few hours each afternoon. Mama was a little worried and made me promise never to leave Sophie's house later than six; but in any case I had no wish to be seen in the streets after sunset. In the beginning Anna joined us, but then her parents found an old teacher who had been a friend of her grandfather's, so her mother thought it would be safer to let take her lessons at home. Sophie and I did not mind. To be quite frank, we were rather pleased.

I grew to look forward to those afternoons in Sophie's place. Sometimes her father would come and work with us and to my great surprise I enjoyed it. I had never liked him much before, he was always very strict, even more so than Mama. He didn't even want Sophie to have friends. He had once said to Mama, friends were a waste of time. He wanted Sophie to work hard and study Pharmacy so that she could take over the business he had built up. But now he seemed different and quite human. He talked to us as if we were already adults and I had a feeling that for the first time, he took notice of me as a person. He would twist the pencil between his slim nervous fingers and say that life had made us grow up quickly. We would have to face many more problems later on and it was better if we accepted it and prepared ourselves for whatever struggle lay ahead of us. When he said this he did not look superior and supercilious as grown ups usually do when they make such pronouncements, but rather sad. Seeing his face now, mellowed by kindness, I felt quite at home in Sophie's dingy little basement room.

It was in the last week of August that fate finally caught up with me and I escaped just by the breadth of a hair. Later it seemed as if during those long hot summer months the trap had been set. I had been a fool to think I could get away so easily. Of course there are people who always get away, like Anna and her parents, but then I don't think I would really like to be like them.

So far I had always been very careful to follow Mama's instructions to leave Sophie's house not later than six (her father usually accompanied me for half the way) but on that particular afternoon (it was a Friday, I think) both her parents were away and Sophie and I used the opportunity to have a little private chat. There were many things we could not discuss in front of her father. First of all we were both rather doubtful about the Convent school. Sophie, who was a Protestant, had been treated even more shabbily than I had been by the Mother Superior and she dreaded the idea of going there in the autumn. She wanted to know how the nuns had behaved during my first few years at school, but my memories were rather vague and not really very helpful. Time passed quickly and when we finally looked at her old alarm clock (one of the few things her parents had been able to save) it was nearly seven.

I said: 'Good heavens! Mama will be furious. I must really run,' and Sophie accompanied me to the next corner so that we could finish our discussion. Before we parted, we again talked for a while until the clock on the church of St. Joseph struck eight. Quickly we said good-bye.

The road was completely empty but at first I was not really afraid. It was summer and the sun had not yet set. If I ran very quickly, I could be home in 20 minutes. The wooden soles of my shoes cluttered over the pavement and I slowed down because the noise made me feel uncomfortable. There was a short cut to our house from Sophie's place, leading straight through Hammer Park and in normal times we had all used it. But since the Russian invasion people had begun to avoid parks and open spaces and Mama had again and again warned me not to go near such a place. At the crossroad I hesitated. Though I did not want to admit it, I was rather afraid of what Mama would say when I came home so late. Maybe she was already looking for me and I did not want to meet her in such a mood in the street. If I ran through the park I could at least save eight minutes. For a moment I thought I heard a car in the distance but then I rebuked myself for getting fanciful and before I had time to dwell on this particular observation, I turned left and ran straight into the park.

I shall never know why I did not hear the car, but before I was able to understand the implications, it was there. It passed me from behind and then stopped in front of me. The door opened and a man in Russian uniform got out.

He was a big man with a pale, puffed up face and he stood quietly besides the car, his hands on his belt, just looking at me. I just stared back at him as if to convince myself that he was real. Then, very slowly, he pushed back his cap from his forehead and moved towards me.

I understood.

I turned round and ran. But I did not get very far. He caught me by the shoulders and pulled me back. I still could not scream. He jerked me around so that I had to face him and then he pushed me to the ground. We both fell. I tried to scratch his face and kick him but it was like fighting against a piece of stone. It did not make any difference at all. He said nothing, not a word. He just breathed as if he had run very fast and there was a sort of glassy look in his eyes. I kept struggling and we rolled over the rough ground and eventually he came to lie on top of me. I felt the weight of his body, something hard pressing against my belly and with a sudden nauseating clearness I realized what it was. I tried to hit him with my right fist but he caught my wrist and twisted my arm back over my head. His face was just above me and the smell of his breath made me feel sick. I tried to struggle but he was so heavy, I could not move at all.

Suddenly I knew everything. I knew what sexual intercourse meant; I knew how horrible it would be. He would enter my body and break something, break it forever. He would be dirty and wet and full of germs. Without shifting his body he began to pull up my skirt with his free hand and I thought: 'I am trapped, trapped, trapped.' It was like the sounds made by a ticking clock, only the clock was I. My left arm slid over the ground searching for some sort of support and then, suddenly, I felt a stone - a long sharp pointed stone. I picked it up and pushed it into his eye.

There was no resistance. The stone sank in as if I had hit butter and something warm and sticky ran over my fingers. The man gave a dreadful scream and his hands flew to his face. It was the first chance I got. With all my strength I pushed his body away, staggered to my feet and ran.

I ran straight through the park, stumbling over stones and roots. The bushes clung to my clothes and I can still remember the sharp stinging pain of the branches whipping my face. A few times I fell down but I did not stop. I just ran and ran. I laughed and cried and all the time I kept rubbing my hand against my skirt to get rid of the repulsive liquid that stuck to it.

I came out of the place at exactly the same place, where many weeks before, I had found the body of 'auntie', but I did not stop. I just kept on running.

When I reached home, Mama and Grandfather were already waiting for me outside our garden gate and as soon as they saw me they opened the door and let me in. In the hall I sat down on the stairs and, still laughing hysterically, I beat my fists against the steps. I just could not stop. Mama and Grandfather stood there and talked but I could not hear them. It was like a silent film. You saw people moving their arms, opening and closing their mouths, but no sound came out. They seemed like puppets, like dolls, like shadows hidden behind a mist, so unimportant, so unreal, so ridiculous - I just could not stop laughing.

I don't know what they discussed but eventually they pulled me to my feet and supporting me as if I were a sick woman they helped me up the stairs. I felt completely exhausted. Even to lift up my feet was an effort. I wanted to sit down and rest but they forced me to walk up the stairs. All the time the walls kept swaying backwards and forwards as if the house was a ship. At times they slipped away from me, making me feel small and lonely in an immeasurable space, and then they came pouncing back without warning as if to crush me. It was horrible and funny at the same time and I wanted to laugh again but it turned into crying.

At last we reached my room. Grandfather put me to bed and Mama brought me a glass of water but I did not take it. I did not want to touch anything. I did not want to drink anything. I felt sick.

Mama stood over me and shook me by the shoulders and shouted: 'What has happened? For heavens sake, what has happened? Stop being hysterical. You must tell us, we can't help you if you don't tell us what has happened.'

It was then that I realized they thought I had been raped. Somehow that made me feel absolutely mad with anger. I felt as if someone had thrown cold water into my face.

I stopped crying and said sullenly: 'Nothing happened. I pushed a stone into his eye. I hope he is dead.'

Mama and Grandfather looked at each other. Their look was meaningful and unclean and I could have screamed with rage, but I had a feeling if I did not pull myself together, they might call somebody, a doctor, or a neighbour, and I simply could not have faced that. I made a tremendous effort to control myself and it helped a little, though it did not prevent them from cross-examining me.

'How did it happen?'

'Where have you been all this time?'

'You didn't talk to anybody on the street, did you?'

And in between they discussed me in whispers: '... her clothes don't seem to be torn ... she has only a few scratches on the face ... No blood between her legs ... Well, perhaps ...'

I did not answer their questions. I was not quite sure what would have happened if I had tried to talk. I was far removed from them, on an island of loneliness and their cold logical adult questions were piercing me like blinding shafts of sunlight. I looked at them with disgust and hatred and kept on shaking my head and in the end I think they believed me.

September

And then the Convent school started. On the first morning we assembled in the courtyard, each class forming a little group of its own. To begin with I did not recognize our class because there were hardly ten girls left and we had been well over twenty before the end of the war. But then I saw Sophie and Anna, they waved to me and I went over to them. We greeted each other, embarrassed, and I would say almost with suspicion, and I think we all found each other looking different. Maria, who had always been slim and delicate, had become enormously fat and her skin had taken on a peculiar yellow colour. Anna took the first opportunity to whisper into my ears that Maria had been raped, three times. I was terrified Maria might overhear us and I said: 'Shut up, Anna, you are a fool.'

Anna looked offended but at least she kept quiet.

We exchanged polite inquiries about each other's well being and asked 'How had it been,' and we all said, 'Oh, not too bad,' and we all knew that we were lying. Then we tried to find out about the girls who had failed to turn up but all we could gather was that they had left town with their parents before the Russians came. Nobody knew anything further.

Eventually a nun came - she was the same who had met Mama and me in the Visitors' room - she clapped her hands like a ballet mistress, or a prison warden, and said in her well-modulated voice: 'Quiet everybody. We shall now proceed to church, filled with proper calm and reference.'

And one group after another filed through the dim corridors into the dark chapel, quietly filling the pews, the younger girls in front and the older ones further back. It was just as it had been in my childhood, many years previously, before I had first heard the name of Hitler. Yet it was not the same because we all had become different.

I watched the little choir boys in their white laced robes, the beautiful exact movements of the priest, the grey clouds of fragrant incense drifting across the altar, and listened to the priest reading the story of the poor being blessed because for them was the kingdom of heaven. And then he began to preach and he preached - about chastity! He told us about some little Italian girl who was now a saint because she had preferred to be killed by the man, who had sinfully desired her, rather than to submit to rape. He said that many girls in Austria had chosen the same fate, and that the church was proud of them. I could not believe my ears; I looked around and saw Maria going deadly pale under her yellow complexion, her fat body slumping forwards in a helpless and ungainly posture. I saw Sophie look cold and stern with a hard expression around her mouth, an expression I had so far only seen on the faces of women very much older. And I saw Anna's smug and self-satisfied smile. And I wanted to jump up and spit at the priest and slap his face and scream. But of course I did nothing, because by then I had already learned my first lesson and I knew that screaming never made any difference at all.

And that was how it went on.

'Do not wear immodest dresses that show your knees.'

'You must have long trousers for gymnastics. A holy father might pass through the courtyard and see your exposed legs. Just think how embarrassed he would be.'

'You are not permitted to wear such outrageous bathing suits.'

'You are not allowed to do this ... you are not allowed to do that ... you are not allowed ... you are not allowed ... not allowed ... not allowed ...'

The more life became safe the more the grown-ups took over again and made their laws. Suddenly they remembered that we were children and that there were things we were not supposed to know. They became touchingly protective.

One day, while going home from school, I saw for the first time 'for adults only' written across a film title. I simply had to stop and look at it and laugh. It was all so funny. I wondered what could possibly be in that film we did not know already. Could it show more than what I had heard the night Henny had died? Could it depict sex more bluntly, more vulgar than what I had experienced that evening in Hammer Park? But that is not all. Suddenly nobody had been a Nazi. A woman who had once reported Mama to the GESTAPO for giving 'anti-German speeches' went around and told everybody how much she had suffered during that 'terrible Hitler time' and everybody listened and agreed.

I just can't understand it. I just don't know what is wrong with them. Can they forget so easily? Have they suddenly become more dishonest or have they always been like this?

Sometimes I feel sick with disgust. Sometimes it seems as if during all those long and deadly summer months life had been more honest. You knew your enemies and you knew your friends. You knew you could be raped, killed or kidnapped, or you could survive. You knew you could starve or you could get food. You knew whom to hate and whom to love. Now it is no longer so simple. There are times when I am filled with a strange nostalgia, a brooding sadness. I almost wish I could go back and live once more through that terrible summer, or at least go far, far way to some place where life is still rough and blunt and never, never come back to people who cheat and lie and pretend and who will try everything to make me one of them.

The best time of the day is the hour I spend on my way to school. The autumn has only just begun. In the morning the streets are touched by thick white mist and at lunchtime, when I go home, the sky is blue and quiet. The leaves are turning yellow and the last butterflies are becoming slow and tired. Sometimes I watch the crows coming from the fields. They fly high, in long formations, and their black bodies touch each other in flight. Then I am almost happy. I forget the people. So far they have not been able to destroy nature. But make no mistake, they will, they will!

(Many years later I was at a conference in India, in the town of Madurai. At dinner, I sat next to a professor from Moscow.

He said: 'Where do you come from originally?'

Uncomfortable, as always, when I am asked this question, I replied: 'Austria.'

'Where exactly in Austria?'

I said: 'Near Vienna.'

'Oh, Vienna!' he said and smiled. 'I was there at the end of the war. We had a marvellous time!' and then he said: 'I am sorry, I should not have said that.'

I said: 'Don't worry; we have all done it to each other. And if we are not careful, we shall do it again.')